Lecture Notes in Computer Science 5763

Commenced Publication in 1973
Founding and Former Series Editors:
Gerhard Goos, Juris Hartmanis, and Jan van Leeuwen

Editorial Board

David Hutchison
 Lancaster University, UK
Takeo Kanade
 Carnegie Mellon University, Pittsburgh, PA, USA
Josef Kittler
 University of Surrey, Guildford, UK
Jon M. Kleinberg
 Cornell University, Ithaca, NY, USA
Alfred Kobsa
 University of California, Irvine, CA, USA
Friedemann Mattern
 ETH Zurich, Switzerland
John C. Mitchell
 Stanford University, CA, USA
Moni Naor
 Weizmann Institute of Science, Rehovot, Israel
Oscar Nierstrasz
 University of Bern, Switzerland
C. Pandu Rangan
 Indian Institute of Technology, Madras, India
Bernhard Steffen
 University of Dortmund, Germany
Madhu Sudan
 Microsoft Research, Cambridge, MA, USA
Demetri Terzopoulos
 University of California, Los Angeles, CA, USA
Doug Tygar
 University of California, Berkeley, CA, USA
Gerhard Weikum
 Max-Planck Institute of Computer Science, Saarbruecken, Germany

M. Ercan Altinsoy   Ute Jekosch
Stephen Brewster (Eds.)

Haptic and Audio Interaction Design

4th International Conference, HAID 2009
Dresden, Germany, September 10-11, 2009
Proceedings

 Springer

Volume Editors

M. Ercan Altinsoy
Ute Jekosch
Dresden University of Technology
Chair of Communication Acoustics
Helmholtzstraße 18, 01062 Dresden, Germany
E-mail: {ercan.altinsoy, ute.jekosch}@tu-dresden.de

Stephen Brewster
University of Glasgow
Department of Computing Science
Glasgow G12 8QQ, UK
E-mail: stephen@dcs.gla.ac.uk

Library of Congress Control Number: 2009932950

CR Subject Classification (1998): H.5.2, H.5, H.4, H.3, H.5.1, I.3.7

LNCS Sublibrary: SL 3 – Information Systems and Application, incl. Internet/Web and HCI

ISSN   0302-9743
ISBN-10   3-642-04075-6 Springer Berlin Heidelberg New York
ISBN-13   978-3-642-04075-7 Springer Berlin Heidelberg New York

This work is subject to copyright. All rights are reserved, whether the whole or part of the material is concerned, specifically the rights of translation, reprinting, re-use of illustrations, recitation, broadcasting, reproduction on microfilms or in any other way, and storage in data banks. Duplication of this publication or parts thereof is permitted only under the provisions of the German Copyright Law of September 9, 1965, in its current version, and permission for use must always be obtained from Springer. Violations are liable to prosecution under the German Copyright Law.

springer.com

© Springer-Verlag Berlin Heidelberg 2009
Printed in Germany

Typesetting: Camera-ready by author, data conversion by Scientific Publishing Services, Chennai, India
Printed on acid-free paper   SPIN: 12742035   06/3180   5 4 3 2 1 0

Preface

"Hearing is a form of touch. Something that's so hard to describe, something that comes, sound that comes to you... You feel it through your body, and, sometimes, it almost hits your face."

Evelyn Glennie, Touch The Sound

The 4th International Workshop on Haptic and Audio Interaction Design was held in September 2009 in Dresden, Germany (the previous meetings were held in Glasgow, Seoul and Jyväskylä). The conference is the flagship event of the community promoting research and scientific progress in the audio and haptic interaction field. The main focus of the HAID workshop series is to bring together haptic and audio researchers and practitioners who share an interest in finding out how the two modalities can be used together, what are the relative contributions of the different sensory modalities to the multimodal percept, and what are the design guidelines of multimodal user interfaces. The research challenges in the area are best approached through user-centered design, empirical studies and the development of novel theoretical frameworks. There is a strong physical relationship between sound and vibration. Based on this physical relationship, there are significant similarities between auditory and haptic perception. The similarities are also observable between audio and haptic user interfaces and actuators. Therefore, the interaction between audio and haptic researchers is promising in developing new products but also understanding unimodal and multimodal perceptual issues.

A total of 17 papers were accepted for HAID 2009, each containing novel work on these human-centric topics. Each paper was peer reviewed at least twice using an esteemed set of leading international figures from both academia and industry, to whom we are grateful for the quality of their reviews, time and patience. We would also like to thank Sebastian Merchel for his excellent support regarding the Online Conference Service. Below the papers are categorized and summarized based on their application and focus.

Haptic Communication and Perception

Touch is a powerful nonverbal communication tool and plays an important role in our daily life. Moll and Sallnäs investigate how the touch modality can be used in haptic interfaces in order for people to communicate and collaborate. Their results show that haptic feedback can convey much more information than just the "feeling" of virtual objects. Besides realistic haptic rendering of objects, haptic feedback can also be used to provide an abstract feedback channel. De Boeck, Vanacken and Coninx investigate the effect of different magnitudes of force feedback on the user's performance in a target acquisition task.

Navigation and Guidance

Recently, audio and haptic feedbacks have been applied in numerous applications for navigation purposes and accurate path tracing. Four papers in this section contribute to this body of research. The effects of force profile, path shape and spring stiffness on the performance of path tracing, in passive haptic guidance systems are evaluated by Zarei-nia, Yang, Irani and Sepehri. Haptic data visualization is a growing research area. Two types of interaction techniques to help the user get an overview of data are presented by Panëels, Roberts and Rodgers. Mc-Gookin, Brewster and Priego introduce the concept of Audio Bubbles—virtual spheres filled with audio that are geocentered on physical landmarks, providing navigational homing information for a user to more easily locate the landmark. A number of different sonification approaches that aim to communicate geometrical data, specifically curve shape and curvature information, of virtual 3-D objects are introduced by Shelley, Alonso, Hollowood, Pettitt, Sharples, Hermes and Kohlrausch.

Visual Impairment

Haptic and audio interaction bring most important benefits in situations where users may be unable to use a visual display. A system which uses audiotactile methods to present features of visual images to blind people was introduced by Dewhurst. De Felice, Attolico and Distante introduce an approach to design a multimodal 3D user interface for the interaction of blind users with VEs. Miao, Köhlmann, Schiewe and Weber describe two existing paper prototyping methods, visual and haptic paper prototyping, and indicate their limitations for blind users and introduce a new approach "tactile paper prototyping" to design haptic user interfaces.

Vibrotactile Feedback and Music

Not only auditory perception, but also the sense of touch plays an important role in music. Vibrotactile actuators and audiotactile interaction give new possibilities for creative design of musical instruments. Havryliv, Geiger, Guertler, Naghdy and Schiemer have selected the carillon as a music instrument for haptic simulation and they demonstrate analytical techniques that permit the accurate simulation of batons of varying force-feedback. A prototype system that supports violinists during musical instrument learning is presented by Grosshauser and Hermann. Mobile music is a new field which uses audiotactile interaction. Reis, Carriço and Duarte introduce a user-centered iterative design of a multi-user mobile music application: The Mobile Percussionist. Sound and vibration perception are always coupled in live music experience. Merchel and Altinsoy have investigated the influence of whole-body vibrations on the perceived quality of conventional audio reproduction systems.

Multimodal User Interfaces: Design and Evaluation

The use of touchscreens, touch panels and touch surfaces is growing rapidly because of their software flexibility as well as space and cost savings. Research into audiotactile interaction for touch screens has grown in recent years. The potential benefits associated with the provision of multimodal feedback via a touch screen on older adults' performance in a demanding dualtask situation are examined by Lee, Poliakoff and Spence. Altinsoy and Merchel have conducted psychophysical experiments to investigate the design and interaction issues of auditory and tactile stimuli for touch-sensitive displays.

Multimodal Gaming

Haptics and audio are indivisibly linked not only with music but also with entertainment. The aim of Koštomaj and Boh's study was to gain a better understanding of how to design a user's physical experience in full-body interactive games. Rath and Bienert describe a recently implemented game on the MacBook platform intended for use as an easily distributable tool for collection of psychophysical data.

Through the work contained in these papers, it is clear that the benefits of auditotactile interaction are very promising for engineers who design multimodal user interfaces or industrial products. At the same time, there are still many fundamental scientific questions in this field which should be investigated.

September 2009

Ercan Altinsoy
Ute Jekosch
Stephen Brewster

Organization

The 4th International Workshop on Haptic and Audio Interaction Design was organized by the Dresden University of Technology (Germany), Chair of Communication Acoustics and the University of Glasgow (UK), Department of Computing Science.

Conference Chairs

Ercan Altinsoy	Dresden University of Technology (Germany) Chair of Communication Acoustics
Ute Jekosch	Dresden University of Technology (Germany) Chair of Communication Acoustics
Stephen Brewster	University of Glasgow, (UK) Department of Computing Science

Program Committee

Farshid Amirabdollahian	University of Hertfordshire, UK
Nick Avis	University of Cardiff, UK
Stephen Barrass	University of Canberra, Australia
Matt-Mouley Bouamrane	University of Manchester, UK
Seungmoon Choi	POSTECH, Korea
Andrew Crossan	University of Glasgow, UK
Abdulmotaleb El Saddik	University of Ottawa, Canada
Cumhur Erkut	Helsinki University of Technology, Finland
Stephen Furner	British Telecommunications Plc., UK
Matti Gröhn	CSC - Scientific Computing, Finland
Thomas Hempel	Siemens Audiolog. Technik GmbH, Germany
Dik J. Hermes	Eindhoven University of Technology, The Netherlands
Steven Hsiao	John Hopkins University, USA
Ali Israr	Rice University, USA
Gunnar Jansson	Uppsala University, Sweden
Johan Kildal	Nokia Ltd, Finland
Roberta Klatzky	Carnegie Mellon University, USA
Ravi Kuber	University of Maryland, USA
William Martins	University of Sydney, Australia
David McGookin	University of Glasgow, UK
Charlotte Magnusson	Lund University, Sweden
Sebastian Merchel	Dresden University of Technology, Germany
Emma Murphy	McGill University, Canada
Michael Miettinen	Suunto Ltd, Finland

Ian Oakley	University of Madeira, Portugal
Sile O'Modhrain	Queen's University Belfast, UK
Antti Pirhonen	University of Jyväskylä, Finland
David Prytherch	Birmingham City University, UK
Matthias Rath	Deutsche Telekom Laboratories, Germany
Roope Raisamo	University of Tampere, Finland
Chris Raymaekers	Hasselt University, Belgium
Hong Z. Tan	Purdue University, USA
Paul Vickers	Northumbria University, UK
Daniel Västfjäll	Chalmers University of Technology, Sweden
Bruce Walker	Georgia Tech, USA
Fredrik Winberg	KTH, Sweden
Wai Yu	Thales Air Defence Ltd, UK

Industrial Sponsors

Immersion Corporation

Table of Contents

Haptic Communication and Perception

Navigation and Guidance

Visual Impairment

Vibrotactile Feedback and Music

Multimodal User Interfaces: Design and Evaluation

Multimodal Gaming

Communicative Functions of Haptic Feedback

Jonas Moll and Eva-Lotta Sallnäs

Royal Institute of Technology
Lindstedtsvägen 5
SE-100 44 Stockholm
{jomol,evalotta}@csc.kth.se

Abstract. In this paper a number of examples are presented of how haptic and auditory feedback can be used for deictic referencing in collaborative virtual environments. Haptic feedback supports getting a shared frame of reference of a common workspace when one person is not sighted and makes haptic deictic referencing possible during navigation and object exploration. Haptic guiding is a broader concept that implies that not only a single action, like a deictic reference, is made but that a whole sequence of temporally connected events are shared, sometimes including deictic referencing. In the examples presented in this paper haptic guiding is used by participants as a way to navigate while at the same time explore details of objects during joint problem solving. Guiding through haptic feedback is shown to substitute verbal navigational instructions to a considerable extent.

1 Introduction

The touch affordances (Gibson, 1966) that humans found of objects in their context over thousands of years of poking around in the nearby surroundings, together with the cooperative quality of human conversation (Clark and Wilkes-Gibbs, 1986), have most probably been important for human development. Being able to explore with your hands and communicate at the same time facilitate coordination of action and learning.

The question then arises; what communicative functions do haptic feedback in it self have? Here, haptic feedback refers to an integration of both kinaesthetic sensing and tactile sensing (Loomis and Lederman, 1986). First of all, a person can look at some one else using a tool or any object and imitate that movement and secondly they can hold on to the same object and through the joint action understand something about the other persons intentions and ways of doing a task. Gestures can show deictic reference (Clark, 2003; Clark, 1996) to i.e. an object, place or person, a skill (pantomimic gesture), specify manner of motion of an object (Fussell et al, 2004) and so on. Traditionally, gestures have been described as individually performed actions but what happens when people do a gesture together holding on to the same object or to a networked pointing device? Is that still gestures or something else, maybe it is guiding or guidance?

Furthermore, what is the reason why a teacher sometimes feels the urge to take the same tool that the student is using and show a movement by holding on to the same

M.E. Altinsoy, U. Jekosch, and S. Brewster (Eds.): HAID 2009, LNCS 5763, pp. 1–10, 2009.
© Springer-Verlag Berlin Heidelberg 2009

tool? Well, first of all the specific procedure can be taught. But something else is also communicated by the touch feedback that has to do with the force something is done with, and the exact direction in which it has to be done. These aspects are only possible or easier to convey by the touch modality compared to vision or hearing. This is also relevant for what researchers refer to as tacid or implicit knowledge. Apart from the above examples, there are other useful functions of haptic feedback for collaboration such as handing off objects, pulling in each end of something or jointly holding on to an object and do a manoeuvre that needs more than one person, such as folding a sheet. In this paper a number of concrete examples have been selected from two earlier studies that illustrate how the touch modality can be used in haptic interfaces in order for people to communicate and collaborate. This paper is theoretical in that it discusses different conceptual phenomena related to communicative aspects of haptic and auditory feedback, based on examples from the two studies that are presented and discussed. A report of the quantitative results and all the details of the methodology regarding the two studies are not presented in this paper. It is also worth noting that the general findings reported here is the result of an inductive analysis that specifically focus on communicative aspects of haptic and auditory feedback.

2 Background

Communication has been defined in many ways that are more or less restricting. The following definition by Cherry (1957) is useful for the kind of communication considered in this article: *"the psychological signals whereby one individual can influence the behaviour of another"*. This definition opens up for investigating communication in a broad sense that also includes communication that is not verbal which many other definitions require. Research about gestures points out that gesturing in itself can be communicative (Clark, 1996) and that signals are the act of creating meaningful signs to others (Clark, 2003). For deaf people this is obvious and for blind and deaf people one useful communication option is sign language conveyed by haptic feedback. Gibson (1979) argued that humans not only perceive the affordances of objects but that also the social behaviour of other beings have affordances. Humans are dynamic and convey complex patterns of behaviour that other humans interpret as affording certain behaviours reciprocally in a continuous fashion. Humans interact with one another and behaviour affords behaviour. Gibson (1979) argued that:

"The perceiving of these mutual affordances is enormously complex, but nonetheless lawful, and it is based on the pickup of the information in touch, sound, odour, taste and ambient light"

Following this line of reasoning, multimodal input of information is important for an accurate understanding of another person's social affordances. Deictic references are important for common ground, that is defined as a state of mutual understanding among conversational participants about the topic at hand (Clark and Brennan, 1991). This is especially true when the focus of interaction is a physical object. Grounding activities aim to provide mechanisms that enable people to establish and maintain common ground (McCarthy et al, 1991; Sallnäs et al, 2007). Deictic references, like "that", "this", "there" is one kind of grounding activity that direct the partner's attention to a specific object. Maintaining common ground is also shown to be much easier when collaborators can make use of this kind of references (Burke and Murphy, 2007). The importance of providing the possibility to gestures and gaze for deictic referencing in collaborative environments has been acknowledged in a number of studies (Cherubini, 2008; Ou et al, 2003; Kirk et al, 2007; Fussell et al, 2004). In neither of these however, have haptic feedback been utilized for gesturing. It has been pointed out that it is not simple deictic referencing that makes collaboration more efficient, by replacing time-consuming and verbose referential descriptions, but that more complex representational gestures are needed together with simple pointing (Fussell et al, 2004). We argue that the need for addressing both deictic referencing and guiding is a result of the previous statement. Guiding is a broader concept that implies that not only a single action like a deictic reference is made but that a whole sequence of temporally connected events are shared, sometimes including deictic referencing. The concept of guiding as opposed to that of guidance is especially appropriate to use in a collaborative situation involving haptic feedback that is reciprocal, like when a guide dog coordinates its movements with a blind person. Guidance is a commonly used word that is very useful in learning situations (Plimmer et al, 2008). Apart from pointing Clark (2003) argue that placing oneself in relation to other persons and the context as well as placing objects is equally essential indicative acts as pointing. The haptic modality is unique in that it is the only modality with which a person can both modify and perceive information simultaneously in a bilateral fashion. We argue that the implication of that for communication is clear. Two persons holding on to the same object can communicate their intention and perceive the other's intentions by haptic feedback almost simultaneously. In this process both pointing and placing acts are made jointly. A number of studies on collaboration have shown that haptic feedback improves task performance and increases perceived presence and the subjective sense of social presence (or togetherness) for different application areas in shared virtual environments (Ho et al, 1998; Basdogan et al, 2000; Durlach and Slater, 2000; Sallnäs et al, 2000; Oakley et al, 2001). In one study the interaction between visually impaired pupils and their teachers was investigated (Plimmer et al, 2008). This study investigated the effects of training handwriting using haptic guidance and audio output to realize a teacher's pen input to the pupil. However, gesturing including deictic referencing has not been specifically addressed in these studies.

3   The Collaborative Environment

The examples presented in this paper are all based on two evaluations. In one of the evaluations, a haptic and visual application that supported group work about geometry was used (figure 1). In the other evaluation, two versions of a haptic and visual application that supported building with boxes were compared of which one version also included auditory feedback (figure 2). All applications build on the same functionality and all versions of the application are developed for collaborative problem solving. In the applications, the collaborative environment is a ceiling-less room seen from above. Two users can interact at the same time in the environments. Both users have a phantom (3 DOF haptic device) each (figure 1) and apart from feeling, picking up and moving around objects they can also feel each others' forces on a joint object when pulling or pushing it and they can also "grasp" each other's proxies and thereby feel pulling or pushing forces. In one version of the application shown in figure 2 the audio cues were; a grasp sound when pushing the button on the Phantom pen in order to lift an object, a collision sound when an object were placed on the ground, a slightly different collision sound when an object were placed on another object and a locate sound that made it possible to know if the other person's proxy were on the left or right side of your own proxy in the virtual environment.

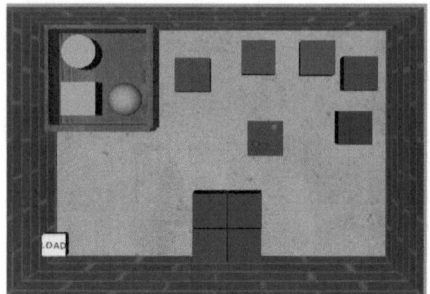 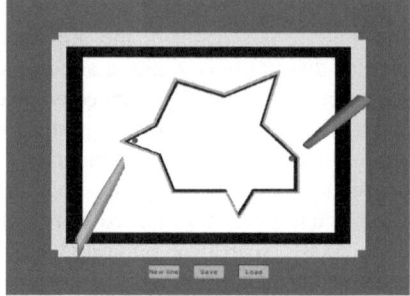

Fig. 1. On the left, two users represented by a blue and red sphere are moving objects in order to cover an area. On the right, two persons are classifying angles using one pen-like proxy each.

Fig. 2. Haptic collaborative application in which both users can pick up and move around objects (small picture in picture). Two users are seen using one haptic feedback device each.

Two evaluations have been performed with this application, which will only be described briefly and related to in the text when needed. In the first study four groups of three pupils (12 in total), of which two were sighted and one was visually impaired, in primary school collaborated in building simple geometrical constructions (Sallnäs et al, 2007). In the second study 14 pairs of sighted and blindfolded university students (28 in total) collaborated in solving mathematically oriented building tasks. Visually impaired people were not recruited to the second study (in which the application version with the audio interface were used) even though it would have been better than blindfolding sighted people. More participants were however needed, than could be found in that target group at that time. In basic research regarding the effects of auditory information on the time to perform two tasks together, it can reasonably be assumed that the overall communicative functions of haptic and auditory feedback are the same for visually impaired and blindfolded sighted people. The general level may be different, but if a parameter has an effect on non-handicapped people, it can be expected to also have an effect on visually impaired people. Especially so, when the context is collaboration between a sighted and a visually impaired person. In that situation, both the communicative frame of reference of the sighted and the visually impaired person have to be taken into account.

4 Communicating Using Haptic Deictic Referencing

4.1 Deictic Referencing as Grounding Strategy in Haptic Interfaces

In the study where pairs of sighted and blindfolded students performed tasks in the application, the potential of the haptic feedback for deictic referencing in the early faces of exploration was shown. In one of the two tasks in particular, where the pair should collaborate in building a cube out of smaller building blocks, deictic referencing was utilized by the blindfolded participant in most of the groups in order to point at objects. The following transcript excerpt illustrates how haptic feedback is used to form a common frame of reference.

Sighted: *Can you feel how big they are?*
                [Blindfolded moves around the cube for a while]
Blindfolded: *These might be 2x2x2*
Sighted: *Yeah, I guess so. But it seems we have five of these 2x2x2 actually,*
                but they are in two different colors, green and blue, so there has to
                be some difference between them
Sighted: *What is the difference?*
Blindfolded: *The height is more on this one*
                [Blindfolded follows the edges of a blue block]
Sighted: *It's high?*
Blindfolded: *Yeah it's longer*

In the above dialogue excerpt the pair is building up a common frame of reference by going through and exploring the different objects present in the scene. Deictic referencing supported by haptic feedback is used in especially two places above, when the collaborating partners want to agree on the dimensions of two different building blocks. When the blindfolded moves the proxy over the edges of the blocks as he says "this one" and "These might be..." he makes it possible for them to talk about and focus the attention on the respective block. Thus, even though the blindfolded did not have access to the visual information on the screen the haptic feedback made it possible for them to build a common frame of reference. The above example shows that the haptic feedback can support direct object referencing in order to create a shared mental model of the workspace. Another typical example of direct object referencing is when the blindfolded indicates a specific object by repeatedly moving up and down with the proxy above it. This is yet another way of focusing the sighted's attention to a specific object of interest. In the same study as referred to in the above excerpt the blindfolded participant often used an indicative act in order to get attention or to place themselves in the work space. One illustrating example, that also says something interesting about the relation to the feeling of presence, is when the sighted participant in one group did not know where the blindfolded person's proxy was. The proxy could not be seen because the blindfolded person was just exploring behind an object that was being built. When asked about his whereabouts, the blindfolded person moved under the object until he felt the end of it after which he moved up and down several times in the virtual room saying "Here I am". Thus, he used the haptic feedback to find out when he was "visible" and in this way he could show where he was. The following dialogue excerpt from the same study illustrates another example of deixis supported by haptic feedback:

Blindfolded:	*I think I know..., if we put the..., eh..., 2x2x3 here*
	[He moves up and down against the floor where he wants it]
Sighted:	*Yeah, ok*
	[Sighted participant puts the block in position
Blindfolded:	*And then the 2x4x1 here...*
	[Blindfolded indicates position by moving back and forth on the floor just under the 2x2x3-block just placed]
Blindfolded:	*And then you put the 2x2x2 here*
	[Blindfolded indicates position by pointing in a similar manner as before]

In this dialogue excerpt the blindfolded person uses both gestures and direct references to places to indicate where and how he wants the sighted peer to put the different building blocks. When telling the sighted peer where to put 2x4x1-block, in the above example, the blindfolded participant uses a gesture when he moves along the floor back and forth to show the intended placement and orientation of the block. In the other cases he uses the haptic device to point to specific locations in the room, to show to the sighted where he wants the different blocks to be put. This example clearly shows how you can use haptic feedback for direct referencing to locations.

4.2 Deictic Referencing by Haptic Guiding

One of the most important findings from the two studies was that different kinds of haptic guiding have a great potential for deictic referencing when it comes to supporting collaboration in haptic interfaces. Two kinds of haptic guiding have been evaluated in the studies. First, two users can hold on to the same object and feel each other's forces on it enabling them to drag each other around. The second kind of haptic guiding function enables a user to actually grab the other one's proxy directly by pushing a button on the phantom when the two proxies get into contact with each other. The following example, from the study with sighted and visually impaired pupils where a visually impaired pupil and a sighted pupil respectively collaborated in solving some simple geometrical tasks, clearly shows the benefit of using haptic guiding for deictic referencing:

Sighted: *All right, you can pick that one you have now*
                  [The visually impaired pupil picks up the cube]
Sighted: *And then, ..., wait, ..., a little bit more to the right.*
                  Up, take it up a little, ...,
                  [The visually impaired pupil moves up towards the roof]
Vis. impaired: *No, it does not work*
                  [The sighted guy picks up the same cube to help]
                  [They move the cube towards them]
                  [They move the cube a tiny bit to the left]
                  [They place the cube to the right of another one]
                  [They fine-tune the cube's position]
Sighted: *That is good!*
                  [The visually impaired and the sighted pupil let go of the cube]

In the above example the pair of pupils tries out different ways of collaborating when placing a cube in a certain position. The sighted pupil starts out by trying to give direction cues verbally that, according to the visually impaired pupil's utterance, did not work very well. You can see from the excerpt that giving verbal guiding can be cumbersome and the verbal guiding part of the example is actually much longer than shown above. However, when giving verbal guiding did not work the sighted pupil grabbed the same object as the visually impaired pupil held and started to give haptic guiding. As the visually impaired pupil held on to the cube her sighted peer dragged it around and placed it in the correct location. In this way the sighted pupil could use haptic guiding to physically help the visually impaired pupil. Another interesting remark that can be made from the above example is that no one is saying anything during the haptic guiding operation. The haptic guiding function replaces the verbal guiding and in several cases it has been shown to save a lot of time and shift the focus from directing the visually impaired pupil to the actual task at hand. The same conclusions could be drawn from both studies. The following dialogue excerpt is yet another example of haptic guiding, taken from the study with sighted and blindfolded students:

                  [Sighted grabs the blindfolded's avatar]
                  [He drags the blindfolded to the beginning of an L-shape, consisting
                  of two 4x2x1 blocks making a 90 degree angle]
Sighted: *Now, here we have an L-shape..*
                  [Sighted drags the blindfolded to the top of the shape]
Sighted: *... this is the top.*
                  [Sighted now drags the blindfolded back and forth on the L-shape's
                  north-southern part a few times]
                  [He then drags the Blindfolded to the east, until the shape ends]

Sighted:	*Ok, and this is the bottom right... and then we have this cube that is taller than the others*
	[He drags blindfolded up and down on a tall block placed beside the L]
Sighted:	*We have another one just like it*

The example above is taken from the early phases of exploration. As preparation the sighted person had organized the two 4x2x1-blocks in an L-shape and he had also put some other blocks beside the L. In order to show the blindfolded the different blocks and their dimensions the blindfolded then grabbed the blindfolded's avatar directly and dragged him around the different building blocks. This example clearly shows the potential for a haptic guiding function for deictic referencing when it comes to collaboration between sighted persons and persons who lack vision.

5 Communicating with Sound

In the experiments with sighted and blindfolded students, sound functions were also evaluated. One of these was a sound that was heard whenever one of the users pushed the button on their respective phantom. The purpose of this, so called, contact sound was to indicate (in stereo) where the sighted person's avatar was in the environment. The auditory feedback was used for dectic referencing in a similar way as when someone knocks on something and says "here" or "this one". The following dialogue example shows this function in use:

Sighted:	*Pick up a new cube*
	[Blindfolded locates a cube on her own]
Blindfolded:	*That one?*
Sighted:	*Yeah...And then you can move here...*
	[Sighted uses sound to show the way]
	[Blindfolded navigates to a place slightly above the intended one]
Sighted:	*Ok, down a bit..., down..., stop*
	[Blindfolded releases]

In the example, the sighted person makes use of the contact sound to give information about in which direction the blindfolded should move and hereby the blindfolded is guided to approximately the right location. As stated earlier, giving verbal direction cues often is quite cumbersome when one of the collaborating partners cannot use their sight. The contact sound cue facilitates the collaboration since the position to go to is indicated by the sound instead. Thus, auditory deixis can be used to replace verbal communication and shift focus to issues more vital for the task at hand.

6 Discussion

The examples presented in this paper clearly show that the haptic feedback can convey so much more information than just the "feeling" of virtual objects. When two people collaborate in a haptic interface, like the one evaluated in our studies, it is evident that the haptic feedback also can be used to communicate both intentions and information much in the line with Cherry's (1957) definition of communication. The examples have shown that the haptic feedback facilitates grounding during the early phases of exploration, especially through the use of deictic referencing as Clark (1996) suggests that gesturing does. We have also shown that added guiding functions can aid the collaboration in a haptic environment considerably, especially when one in the team lacks sight that has also been shown by Plimmer et al. (2008). By using haptic guiding (holding on to the same object or dragging the other user's avatar directly) one can communicate information about direction that does not need to be verbalized. It is shown in the examples that collaboration and joint problem solving becomes much easier when you do not have to focus on constantly giving direction cues. The examples used in this paper are all based on studies where one of the collaborating partners has been blindfolded or visually impaired. However, much of the findings reported here could surely be applied to pairs of sighted persons as well.

Acknowledgements

The Swedish Research Council that funds the MCE project is greatly appreciated.

References

Basdogan, C., Ho, C., Srinivasan, M.A., Slater, M.: An Experimental Study on the Role of Touch in Shared Virtual Environments. ACM TOCHI 7(4), 443–460 (2000)

Burke, J., Murphy, R.: RSPV: An Investigation of Remote Shared Visual Presence as Common Ground for Human-Robot Teams. In: HRI 2007, Virginia, March 8-11, pp. 161–168 (2007)

Cherry, C.: On human communication. MIT Press, Cambridge (1957)

Cherubini, M., Nüssli, M.-A., Dillenbourg, P.: Deixis and gaze in collaborative work at a distance (over a shared map): a computational model to detect misunderstandings. In: ETRA 2008, pp. 173–180. ACM, New York (2008)

Clark, H.: Pointing and Placing. In: Kita, S. (ed.) Pointing: Where Language, Culture, and Cognition Meet, pp. 243–268. Lawrence Erlbaum Associates, Mahwah (2003)

Clark, H.H.: Using language. Cambridge University Press, Cambridge (1996)

Clark, H.H., Brennan, S.E.: Grounding in Communication. In: Resnick, L., Levine, J., Teasley, S. (eds.) Perspectives on Socially Shared Cognition, pp. 127–149. American Psychological Association, Hyattsville (1991)

Clark, H.H., Wilkes-Gibbs, D.: Referring as a collaborative process. Cognition 22, 1–39 (1986)

Durlach, N., Slater, M.: Presence in shared virtual environments and virtual togetherness. Journal of Presence: Teleoperators and Virtual Environments 9(2), 214–217 (2000)

Fussell, S.R., Setlock, L.D., Yang, J., Ou, J., Mauer, E.M., Kramer, A.: Gestures over video streams to support remote collaboration on physical tasks. Human-Computer Interaction 19, 273–309 (2004)

Gibson, J.J.: The senses considered as perceptual systems. Mifflin, Boston (1966)

Gibson, J.J.: The ecological approach to visual perception. Mifflin, Boston (1979)

Ho, C., Basdogan, C., Slater, M., Durlach, N., Srinivasan, M.A.: An experiment on the influence of haptic communication on the sense of being together. In: Proceedings of the British Telecom Workshop on Presence in Shared Virtual Environments (1998)

Kirk, D., Rodden, T., Fraser, D.S.: Turn it This Way: Grounding Collaborative Action with Remote Gestures. In: CHI 2007, San Jose, California, April 28-May 3, pp. 1039–1048 (2007)

Loomis, J.M., Lederman, S.J.: Tactual perception. In: Boff, K.R., Kaufman, L., Thomas, J.P. (eds.) Handbook of perception and human performance, pp. 31.31-31.41. Wiley/Interscience, New York (1986)

McCarthy, J., Miles, V., Monk, A.: An experimental study of common ground in text-based communication. In: Proceedings of the SIGCHI conference on Human factors in computing systems: Reaching through technology, pp. 209–215. ACM Press, New York (1991)

Oakley, I., Brewster, S.A., Gray, P.D.: Can You Feel the Force? An Investigation of Haptic Collaboration in Shared Editors. In: Proceeding of Eurohaptics 2001, pp. 54–59 (2001)

Ou, J., Fussell, S.R., Chen, X., Setlock, L.D., Yang, J.: Gestural Communication over Video Stream: Supporting Multimodal Interaction for Remote Collaborative Physical Tasks. In: ICMI 2003, pp. 107–114 (2003)

Plimmer, B., Crossan, A., Brewster, S., Blagojevic, R.: Multimodal Collaborative Handwriting Training for Visually-Impaired People. In: Proceedings of CHI 2008. ACM Press, New York (2008)

Sallnäs, E.-L., Moll, J., Severinson-Eklund, K.: Group Work about Geometrical Concepts Including Blind and Sighted Pupils. In: World Haptics 2007, Tsukuba, Japan, pp. 330–335 (2007)

Sallnäs, E.-L., Rassmus-Gröhn, K., Sjöström, C.: Supporting presence in collaborative environments by haptic force feedback. ACM TOCHI 7(4), 461–476 (2000)

Target Acquisition with Force Feedback: The Effect of Different Forces on the User's Performance

Joan De Boeck, Lode Vanacken, and Karin Coninx

Hasselt University, Expertise Centre for Digital Media (EDM)
and transnationale Universiteit Limburg
Wetenschapspark 2, B-3590 Diepenbeek, Belgium
{joan.deboeck,lode.vanacken,karin.coninx}@uhasselt.be

Abstract. Besides realistic haptic rendering of objects, haptic feedback can also be used to provide an abstract feedback channel. This can either be realised by a tactile or a force feedback stimulus. When using forces, care has to be taken that the user's performance is not influenced in a negative way. However, as it is not obvious to determine a suitable force, and currently not many guidelines exist. Therefore, in this paper we investigate the influence on some important parameters that define a force (shape, duration and amplitude). In order to compare different forces, we propose to use the definite integral (Force Integral, *FI*) which combines the considered parameters. From the conducted experiment we learn that the *FI* can be used (within bounds) to make an estimation of the result of the force. Besides this, we also found that above a given *FI* value, the user's performance degrades significantly.

1 Introduction

Over the last decade haptic feedback, which exploits our human sense of touch, has been gaining importance. Haptic feedback, in the form of tactile feedback or force feedback [1] has the ability to provide a very direct feedback loop closely coupled with the user's action (motion). Practical applications applying force feedback often try to render the generated forces as realistic as possible, to provide the user with a natural sensation. Several simulation applications may serve as examples of this approach [2].

In contrast with the generation of realistic forces, other research focuses rather on the *extra informational channel* provided by haptic feedback, independent of the realism of the generated forces. This extra feedback provides users with additional information during their interaction as such that their experience or their performance can be improved. Very often, examples of this kind of feedback can be found in the domain of tactile feedback, including 'Tactons' [3] or the use of a haptic belt for navigation at sea [4]. This extra haptic channel can also be established by using abstract force feedback, e.g. to support pointing tasks in a desktop application or in a virtual environment.

This support can be achieved by *assisting* the user, e.g. using 'gravity wells' (a 'snap-to' effect that pulls the user's pointer to the centre of the target) [1,5]. Using gravity wells showed small non-significant improvements in time, while reducing the error rate.

M.E. Altinsoy, U. Jekosch, and S. Brewster (Eds.): HAID 2009, LNCS 5763, pp. 11–20, 2009.
© Springer-Verlag Berlin Heidelberg 2009

 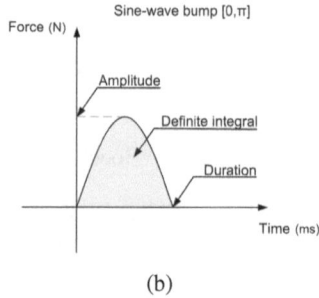

(a)   (b)

Fig. 1. (a) Setup of the experiments: Phantom haptic device and the ISO 9241-9 tapping circle on the screen. (b) An illustration of the force evolution of a sinusoidal haptic bump over time. The grey shaded area is the definite integral, which in this paper we refer to as the force integral (*FI*).

However, in more complex situations where distractor targets are present, gravity wells become disturbing [6,7].

Alternatively, forces can be applied by giving *feedback* in the form of a bump or a vibration as soon as an event occurs. Akamatsu et al. [8] and Cockburn et al. [9] used tactile vibrations to indicate when the mouse cursor hovered a target. They both found that tactile vibration could improve the performance in certain situations although they reported that the vibration could make users miss small targets. In the context of a virtual environment, Vanacken et al. [10] applied a small sine-wave force of a short duration to indicate when a user had switched between indicated targets. No significant performance benefits were found, but small improvements in time were seen.

Force feedback that assists the user, such as gravity wells, can be detrimental and has no guarantee to be beneficial for the performance of the user. On the other hand, force feedback used as extra feedback has showed possible beneficial possibilities in user interfaces but it is unclear how to assure the design of the feedback does not have a negative outcome on the performance of the user. Currently not many guidelines exist to support designers in defining the parameters, this implies that the design is mostly performed using 'trial-and-error' until suitable values are found.

Inspired by this observation, the aforementioned research and a first pilot study [11], the work presented in this paper investigates what force strengths, shapes and durations may be useful. In other words, we study how the force parameters influence the user's performance. We propose a rule of thumb that can be applied to combine the aforementioned parameters. This will finally allow us to formulate a guideline concerning the forces that may or may not be applied.

In this paper we define a 'force bump' as a short force with a given *duration* and *amplitude*. The amplitude of the force over time may follow a mathematical pattern such as a sine or a step function, which we define as the force *shape*. Figure 1(b) illustrates a sinusoidal bump. There are more possible parameters that may have an influence on the result of the force-feedback stimulus. Other parameters such as force directions, or device differences will be covered in future work.

To combine the parameters taken into account, we formulate a hypothesis using the definite integral of the force bump (see Figure 1(b)). This allows us to predict the influence of other force bumps with other amplitudes, shapes and durations. We verify the validity of this hypothesis in the following experiment. This hypothesis and the resulting guideline are a valuable tool for designers who want to apply force feedback for a targeting, pointing or a selection task. They can use these results to know what force values must not be exceeded.

2   Force Shape and Duration Experiment

To provide a more convenient calculation of the different parameters (shape, amplitude and duration), we apply the definite integral (Figure 1(b)). Derived from the physical relationship between the applied force and the velocity and position (where position is the double integral of the force), we hypothesise that the definite integral of the force bump may be a decent prediction of its influence on the user's behaviour. In what follows, we define the force integral (*FI*) as $\int_0^T abs(A \cdot f(t))dt$.

In this section we try to verify this assumption by measuring the results of haptic bumps with different force shapes, different durations and different amplitudes, but with the same *FI*, and compare how they relate to each other.

2.1   Apparatus

The experimental setup consisted of a regular 19-inch CRT monitor and a Phantom premium 1.0 haptic device (see Figure 1(a)). The control display gain was 1, which means that one cm physically moved with the device, corresponds to one cm on the screen. Calibrating the device learned us that forces provided by the software were nearly equal to the final forces measured at the device. However gravity compensation (unbalanced weight [12]) of 0.08N downward was required in software.

Another important factor that had to be taken into account, was the inertia and the internal friction of the device. Obviously, we desire values that are as low as possible because a higher friction and higher inertia may interfere with the haptic bump (e.g. 'smooth out' the force bump). As the Phantom premium is designed to keep these values as small as possible, this device suits our needs.

2.2   Participants

Ten participants (two females and eight males) served as participants in this experiment. Participants were selected among co-workers and had an age between 25 and 31 years old with an average of 27. All participants except one were right-handed and used their dominant hand during the experiment.

2.3   Procedure

A simple multidirectional point-select task, as described in ISO 9241-9 [13], was used for this experiment. Ten targets were placed in a circle on the screen (see Figure 1(a)). The diameter of the circle was determined at 6 cm and the size of a target at 0.7 cm (we use physical measures rather than pixels, since pixel sizes vary from display to display). This task has a Fitts' index of difficulty of 3.26 bits, a measure typically used in Fitts' law experiments to indicate the difficulty of the task [14]. This value is chosen to be comparable to the task difficulty of a typical icon selection task [15,16].

We also took into account the implications of the movement scale; the limb segments of the user involved in the task depend on the physical distance that has to be covered [17,18]. Usually, the operation of desktop haptic devices is situated in the range of the wrist and fingers. Therefore a 6 cm distance appeared to be a good value [19], as it will adhere to typical movements to be expected with the device.

During the test, the ten targets were highlighted one after the other and users were requested to select (by pointing and clicking) the highlighted target as efficient (fast and accurate) as possible. Highlighting is altered between opposite sides of the circle so that it requires the user to make movements equally distributed among all directions with a maximum distance between the targets.

As the task to perform was a 2D selection task and the haptic device we used is a 3D input device, a vertical guiding plane restricted the task to two dimensions. In order to make sure that users did not use the guiding plane as extra support, as such that the forces had less impact on their movement, we provided them with extra visual feedback about their position inside the guiding plane. The background colour was completely black within a certain offset of the guiding plane and faded to white the more the user pushed into the plane. Users were instructed to avoid having a grey/white background.

Finally, force feedback appearing in the form of a force bump with given shape, duration and amplitude was activated at exactly half-way in the path to the next target. Note that in this basic experiment, the forces serve as a distractor without beneficial goal. This experimental approach should allow us to investigate the user's performance when different forces are applied.

2.4   Independent Variables

As mentioned before, in this experiment we investigate how the user's performance is influenced by the applied forces when parameters such as force shape (S), duration (T) and amplitude (A) are altered. These parameters are all combined in the *FI*.
We consider the following shapes with the following duration:

- A sine wave over half a period, 75 milliseconds ($S=sin_{[0,\pi]}$, $T=75ms$)
- A step function, 40 milliseconds ($S=sqr$, $T=40ms$)
- A sine wave but with a longer duration, 110 milliseconds ($S=sin_{[0,\pi]}$, $T=110ms$)
- A full sine wave $[0,2\pi]$ of 75 milliseconds. It is interesting to see how this shape will behave as it produces positive and negative forces ($S=sin_{[0,2\pi]}$, $T=75ms$)

The amplitudes of the different forces are chosen as shown in Table 1, so that the *FI* of each nth line is the same.

The force integral values ($FI = 0.0$, 9.55, 19.10, 28.65, 38.20, 47.75, 57.30, 66.85, 76.39, 85.94 and 95.49) and the shapes S ($sin_{[0,\pi]}$, $T=75ms$; sqr, $T=40ms$; $sin_{[0,\pi]}$, $T=110ms$ and $sin_{[0,2\pi]}$, $T=75ms$) serve as the independent variables during the design and analysis of this experiment.

Table 1. Equivalent force amplitudes calculated using the force integral value and the definite integral

Force Integral	$\sin_{[0,\pi]}$ (75ms)	sqr (40ms)	$\sin_{[0,\pi]}$ (110ms)	$\sin_{[0,2\pi]}$ (75ms)
0.0	0.0N	0.0N	0.0N	0.0N
9.55	0.2N	0.24N	0.14N	0.2N
19.10	0.4N	0.48N	0.27N	0.4N
28.65	0.6N	0.72N	0.41N	0.6N
38.20	0.8N	0.96N	0.55N	0.8N
47.75	1.0N	1.19N	0.68N	1.0N
57.30	1.2N	1.43N	0.82N	1.2N
66.85	1.4N	1.67N	0.95N	1.4N
76.39	1.6N	1.91N	1.09N	1.6N
85.94	1.8N	2.15N	1.23N	1.8N
95.49	2.0N	2.39N	1.36N	2.0N

2.5 Design

A repeated measures within-participant design was used. The independent variables *FI* and *S* combined in a fully crossed design resulted in 44 combinations in total (11 *FI*s and 4 *S*s).

Each participant performed the experiment in one session lasting about 25 minutes. The session consisted of 5 blocks with each block containing the 44 combinations repeated 3 times in a random order. For a total of 132 trials per block, this resulted in 660 trials per participant. Between each block, users were obliged to take a break of at least 15 seconds to minimise fatigue during the test. Before starting the experiment, participants were given all 44 conditions in random order to familiarise them with the task. During the experiment, the time it took to select a target was recorded, as well as the amount of errors made during selection.

2.6 Results

In a first step, we investigated the general learning effect during our experiment by comparing the results of the different blocks based on the trial completion time. As a result, the first two blocks were removed to eliminate the results of any learning effect (*Block* ($F_{4,36} = 8.7$, p < .0001)).

Trial Completion Time. A repeated measures analysis of variance of the faultless selection trials showed no main effect for *S* ($F_{3,27} = 1.92$, p = .151) which implies that the shapes did not differ significantly from each other in trial completion time: 881.6ms for $sin_{[0,\pi]}$, *T=75ms*, 893.2ms for *sqr, T=40ms*, 874.9ms for $sin_{[0,\pi]}$, *T=110ms* and 907.5ms for $sin_{[0,2\pi]}$, *T=75ms*. This result was to be expected, as we hypothesised that the *FI* values would be the most important factor with regard to the trial completion time of the user. Analysis showed a main effect for *FI* ($F_{10,90} = 8.6$, p < .0001). Post hoc comparisons showed that trial completion time slightly (but non-significantly) deteriorates

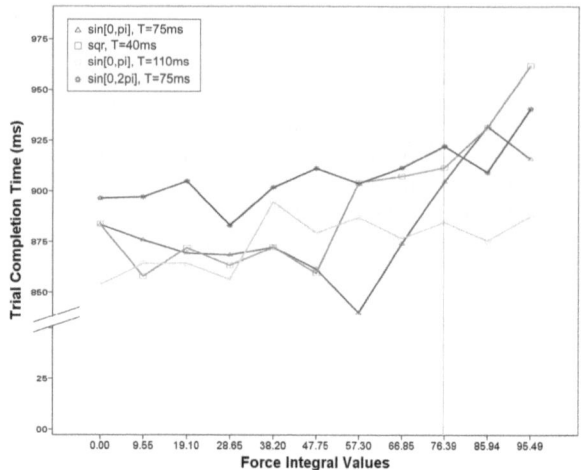

Fig. 2. Force integral values by shape. Significant deterioration is found for FI-values greater than 76.39.

while the force increases. This pattern is true until a certain *FI*-value (*FI=76.39*) from where the trial completion time deteriorates significantly (p < 0.015). Although the trial completion time penalty is small in absolute values, nevertheless it is significant; and moreover, as selection is such a common and frequently used action, even a deterioration of a small amount of time may induce user annoyance.

Although *S* did not show a significant main effect, it did show an interaction effect with *FI* ($F_{30,270} = 1.66$, p < .02). Figure 2 shows the interaction: all shapes show a similar pattern with regard to the force integral, except for $sin_{[0,\pi]}$, *T=110ms*. This shape does not seem to have an equally strong deterioration at the higher *FI* conditions. Several reasons may cause this effect, but future research is necessary to verify our suppositions: it can be argued that the less pronounced deterioration of $sin_{[0,\pi]}$, *T=110ms* is due to the lower amplitude which may be partly masked by the friction of the device. Alternatively, it can also be argued that the longer period of force activations gives more opportunity for the user's reflexes to counter the deviation and apply a compensation.

Velocity Analysis. The study of velocity profiles can provide us with a deeper understanding of the different stadia in the user's motion. Figure 3 shows a typical lateral (top graphs in each figure) and longitudinal (bottom graphs in each figure) velocity profile for different shapes for different *FI*s. We have to stress that the graphs shown in this figure are individual selection trials of individual users. It has to be noted that the entire population of all velocity profiles is subject to a large variation. However, the selected graphs give a good representation of the velocity's behaviour in general.

Figure 3(a) shows a selection without applied force. In the topmost graph, we see a small lateral velocity variance around zero. The longitudinal velocity behaves according to the optimised initial impulse model of Meyer et al. [20]. The ballistic movement (BM) and controlled movement (CM) are indicated in Figure 3(a).

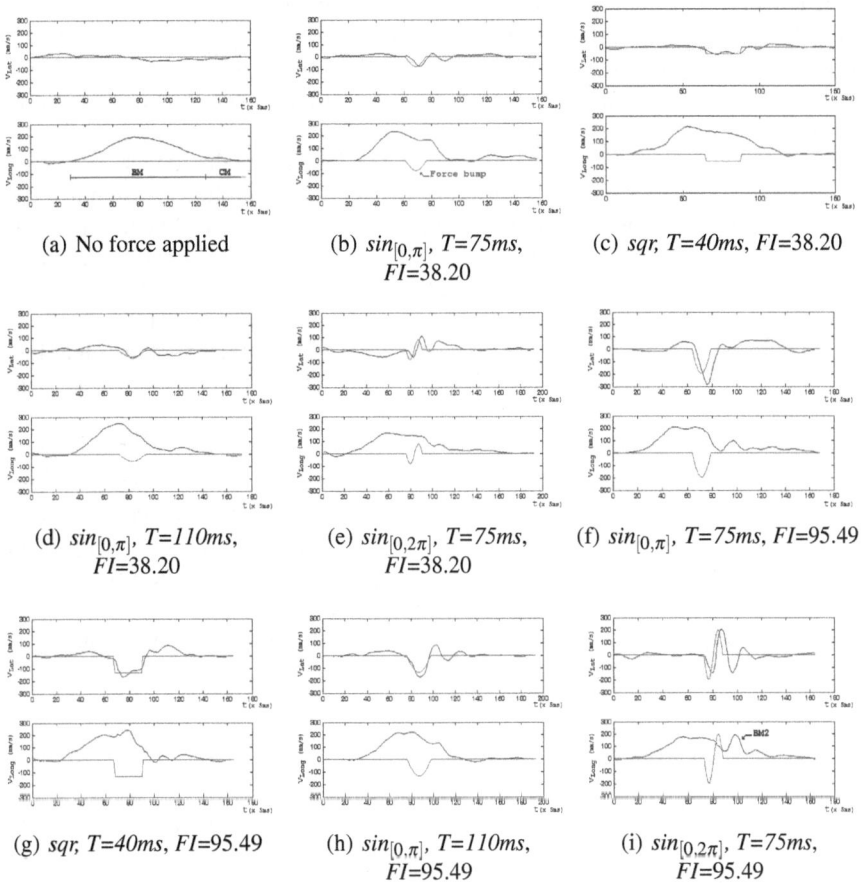

Fig. 3. Velocity Profiles of some 'typical' movements: Upper graphs contains the lateral velocity (orthogonal with moving direction), lower graphs contain the longitudinal velocity (parallel with moving direction)

Looking at Figures 3(b), 3(c), 3(d) and 3(e), they represent the velocity profiles when a small force (FI=38.20) for the respective shapes is applied. In all cases, we see a clear influence of the bump on the lateral velocity, in the form of a small oscillation. From the longitudinal velocity profile, we can learn that it is not (or only slightly) affected by the force bump.

Figures 3(f), 3(g), 3(h) and 3(i) show the velocities when a large force (FI=95.49) is applied. Obviously, we see a similar but larger effect on the lateral velocity. Surprisingly however, the behaviour of the longitudinal velocity, drops down to zero immediately after the bump. This dramatic decrease of velocity is not intended as the end of the ballistic movement phase, which is confirmed by many profiles that show a new (second) but shorter ballistic movement phase (BM2 in Figure 3(i)). Manually categorizing the profiles shows that 80% of the trials ($sin_{[0,\pi]}$, T=75ms, FI=95.49) have a clear speed reduction, or even a complete halt. Probably the temporarily 'halt' induced by the user will help to dampen the unwanted oscillation caused by the force bump. We believe that the significant deterioration discussed in the previous sections is mainly caused by this (unconscious) speed reduction, rather than by the extra distance physically induced by the oscillation.

From the analysis in the previous section, we found that the longer sine ($sin_{[0,\pi]}$, $T=110ms$) behaved in a somewhat different manner. It did not show the significant trial completion time deterioration we should expect for the highest two FI conditions. From the analysis of the velocity profiles, we found that only 64% of the selection trials had a clear speed reduction or halt (compared to 80% for $sin_{[0,\pi]}$, $T=75ms$). Figure 3(h) shows how the longitudinal velocity is less affected, with only a short speed reduction and no halt, as shown in Figure 3(g). This makes us favour the hypothesis that longer bumps give more opportunity for the user to compensate, so that (involuntarily) halting the movement to dampen the oscillation is less necessary.

Error Rates. To analyse the behaviour of the error rate, an error was defined as when the user misclicked the correct target. The overall error rate for the experiment was 139 errors or 2.1%. The shape had no significant effect on the error rate S ($F_{3,27} = .657$, p $= .586$). The force integral values showed no significant effect on the error rate either, although the p-value approached the significance level FI ($F_{10,90} = 1.78$, p $= .076$). After the experiment, several users reported involuntary miss-clicks in case of the large force conditions, due to these forces.

3   Discussion and Implications

Our experiment investigated the user's performance when different forces are applied during a simple multidirectional point-select task.

As expected, the shape, duration and amplitude of a force bump have influence on the user. The number of parameters, however make it difficult to compare the different forces. As a guideline we proposed to express the 'size' of a force bump by means of the force integral (FI), which can be visualised by the area below the force graph (see Figure 1(b)). Even though we do not claim that different force shapes are similar with respect to the user's performance, they do not differ significantly. This allows a designer to use the FI as a good estimation of result of a force bump.

We found that above a certain force integral value, the trial completion time significantly deteriorates. From our experiment in particular, we could deduce that designing abstract force feedback with forces below a FI-value of 76.39 will not be harmful for the user's performance. For the larger force integral values, we could learn from motion path analysis that the longitudinal velocity dropped detrimental immediately after the force bump. We believe that this involuntary halt is the main cause of the user's performance penalty.

It may be clear however that FI is only a rule of thumb, which will have its limits in practice. Although the exact limits must be deduced in future experiments, we found that the longer sine wave ($sin_{[0,\pi]}$, $T=110ms$) behaves somewhat different. This can be intuitively understood from the physical properties of the human-device system.

Very short (strong) bumps ultimately will loose their effect because of the inertia, while similarly, very long (but soft) bumps will disappear by the friction of the device, or in extremis are below the user's 'just noticeable difference [21]'. But other effects such as the longer forces that are easier to compensate by the user (as could be seen from the velocity analysis) may play an equally important role, as well.

We believe that the results of this experiment have a practical benefit. As stated in the introduction, Vanacken et al. [10] use small sinusoidal forces to indicate when the user has switched to another indicated target. With a half sine wave of 25ms and peak value of 1N ($sin_{[0,\pi]}$, $T=25ms$, $A=1N$), resulting in a force integral of $FI=15.92$, this is below the threshold we found. Hence the force should not have any significant deteriorating effect on the user's movement, which is confirmed by the results as the authors found a small (but non-significant) improvement when force feedback was enabled.

Although our hypothesis was tested in a selection task with a constant Fitts' index of difficulty, it can also be used in other types of interfaces in which motion is involved, such as crossing based interfaces [22] or gesture interfaces [23]. Finally, Cockburn et al. [9] found that large amplitudes of tactile feedback may cause users missing small targets. We believe that in this context our approach might be applicable as a rule of thumb for tactile feedback, as well.

4   Conclusions and Future Work

We studied the effect of different magnitudes of force feedback on the user's performance in a target acquisition task. In order to facilitate a comparison of different forces across different parameters such as force shape, duration and amplitude, we proposed to use the definite integral (or Force Integral, FI). We found that the FI can be considered as a good guideline to predict the user's performance. During the experiment, we also observed that above a certain force magnitude the user's performance significantly deteriorates.

The value of this work is to provide user interface designers with a guideline to keep the calculated force integral (based upon the force shape, duration and amplitude) below the force integral values that objectively caused the performance penalty. It is important to note that the results of this investigation do not imply that force feedback below the values found in these experiments is a priori useful. *If* and *when* force feedback can be applied, is still up to the designer to decide.

We found that the FI rule of thumb is a good approximation *within bounds*. In future work, the extremes to which the prediction is valid should be defined more accurately. Another interesting question is whether the 'Fitts' index of difficulty' will have an influence on the results of these experiments. Finally, the Phantom haptic device was used because it has a low inertia and friction. Using other devices such as the popular Novint's Falcon, might have an implication with regard to our findings.

Acknowledgements

Part of the research at EDM is funded by the ERDF (European Regional Development Fund) and the Flemish government.

References

1. Oakley, I., McGee, M.R., Brewster, S., Gray, P.: Putting the feel in 'look and feel'. In: CHI 2000, pp. 415–422 (2000)
2. Stone, R.J.: Haptic feedback: A brief history from telepresence to virtual reality. In: Brewster, S., Murray-Smith, R. (eds.) Haptic HCI 2000. LNCS, vol. 2058, pp. 1–8. Springer, Heidelberg (2001)
3. Brewster, S., Brown, L.: Tactons: Structured tactile messages for non-visual information display. In: AUIC 2004, pp. 15–23 (2004)
4. Van Erp, J., Jansen, C., Dobbins, T., Van Veen, H.: Vibrotactile waypoint navigation at sea and in the air: two case studies. In: Eurohaptics 2004, Munich, Germany (2004)
5. Wall, S., Paynter, K., Shillito, M., Wright, M., Scali, S.: The effect of haptic feedback and stereo graphics in a 3d target acquisition task. In: Eurohaptics 2002, Edinburgh, UK (2002)
6. Ahlström, D., Hitz, M., Leitner, G.: An evaluation of sticky and force enhanced targets in multi target situations. In: NordiCHI 2006, pp. 58–67 (2006)
7. Hwang, F., Langdon, P., Keates, S., Clarkson, J.: The effect of multiple haptic distractors on the performance of motion-impaired users. In: Eurohaptics 2003, pp. 14–25 (2003)
8. Akamatsu, M., MacKenzie, I.S., Hasbrouc, T.: A comparison of tactile, auditory, and visual feedback in a pointing task using a mouse-type device. Ergonomics 38, 816–827 (1995)
9. Cockburn, A., Brewster, S.: Multimodal feedback for the acquisition of small targets. Ergonomics 48, 1129–1150 (2005)
10. Vanacken, L., Grossman, T., Coninx, K.: Multimodal selection techniques for dense and occluded 3d virtual environments. International Journal on Human Computer Studies 67, 237–255 (2009)
11. Vanacken, L., De Boeck, J., Coninx, K.: Force feedback magnitude effects on user's performance during target acquisition: a pilot study. Accepted for Interact 2009 (2009)
12. Massie, T.H., Salisburg, J.K.: The PHANToM haptic interface: A device for probing virtual objects. In: ASME 1994, pp. 295–302 (1994)
13. ISO: ISO/TC 159/SC4/WG3 N147. Ergonomic requirements for office work with visual display terminals (VDTs) - Part 9 - Requirements for non-keyboard input devices. (May 25, 1998)
14. Fitts, P.: The information capacity of the human motor system in controlling the amplitude of movement. Journal of Experimental Psychology 47, 381–391 (1954)
15. Douglas, S.A., Kirkpatrick, A.E., MacKenzie, I.S.: Testing pointing device performance and user assessment with the iso 9241, part 9 standard. In: CHI 1999, pp. 215–222 (1999)
16. Soukoreff, R.W., MacKenzie, I.S.: Towards a standard for pointing device evaluation, perspectives on 27 years of fitts' law research in hci. Int. J. Hum.-Comput. Stud. 61, 751–789 (2004)
17. Langolf, G., Chaffin, D., Foulke, J.: An investigation of Fitts' law using a wide range of movement amplitudes. Journal of Motor Behavior 8, 113–128 (1976)
18. Balakrishnan, R., MacKenzie, I.S.: Performance differences in the fingers, wrist, and forearm in computer input control. In: CHI 1997, pp. 303–310 (1997)
19. Accot, J., Zhai, S.: Scale effects in steering law tasks. In: CHI 2001, pp. 1–8 (2001)
20. Meyer, D., Abrams, R., Kornblum, S., Wright, C., Smith, J.: Optimality in human motor performance: Ideal control of rapid aiming movements, 340–370 (1988)
21. Tan, H.Z., Srinivasan, M.A., Reed, C.M., Durlach, N.I.: Discrimination and identification of finger joint-angle position using active motion. ACM Trans. Appl. Percept. 4, 10 (2007)
22. Accot, J., Zhai, S.: More than dotting the i's — foundations for crossing-based interfaces. In: CHI 2002, pp. 73–80 (2002)
23. Bau, O., Mackay, W.: Octopocus: A dynamic guide for learning gesture-based command sets. In: UIST 2008 (2008)

Evaluating Factors that Influence Path Tracing with Passive Haptic Guidance

Kurosh Zarei-nia[1], Xing-Dong Yang[2], Pourang Irani[3], and Nariman Sepehri[1]

[1] Dept. of Mechanical Engineering, University of Manitoba, Winnipeg, MB, Canada
[2] Dept. of Computing Science, University of Alberta, Edmonton, AB, Canada
[3] Dept. of Computer Science, University of Manitoba, Winnipeg, MB, Canada
`umzarein@cc.umanitoba.ca`, `xingdong@cs.ualberta.ca`,
`irani@cs.umanitoba.ca`, `sepehri@cc.umanitoba.ca`

Abstract. A very common task in medical applications and motor-skill training is to trace a path. However, when designing a haptically guided interface, designers need to consider the choice of several parameters in the design. These include the real-time function for bringing back the user to the right path, the effect of the path's curvature on tracing, and the amount of haptic force needed for guiding the user appropriately. In this paper, we describe the results of an experiment that was designed to assess the effect of several design factors that can influence the performance of path tracing tasks. Our results show that the shape of the path has an effect on the amount of deviation from a path. Additionally, we found that a high amount of stiffness is preferred over low stiffness. Finally, the type of force profile that haptically guides the user, particularly the slope of the function, is also an important factor in path tracing tasks. We discuss our results with implications for designs of systems necessitating haptic force feedback in constrained path tracing tasks.

Keywords: Haptic guidance, haptic interface, motor skill training, force feedback.

1 Introduction

Tracing a path accurately is important in many tasks including medical applications and motor-skill training. Medical applications such as tele-ultrasound scanning and bronchoscopy require an operator to guide a probe along a defined path obtained from one or more video streams. Significant deviations from the path (this happens when an ultrasound probe is lifted too far off the patient's body) during a trace will result in incoherent images or can harm the patient (both ultrasound and bronchoscopy). Motor-skill learning is contingent on a number of factors: motivation, practice, learning strategies, and guidance techniques. However, physical guidance is of particular importance because if a learner is unable to trace a trajectory and stay on the path, he/she would not learn the properties of the trajectory in the first place, negatively impacting learning regardless of motivation levels, amount of practice or any other factor. It would be unrealistic to expect any gains in learning outcomes if the guidance was flawed.

M.E. Altinsoy, U. Jekosch, and S. Brewster (Eds.): HAID 2009, LNCS 5763, pp. 21–30, 2009.
© Springer-Verlag Berlin Heidelberg 2009

Recently, haptic-guidance systems have been widely used in facilitating precise path tracing in medical applications [13, 14, 15] and in motor-skill training systems [2, 4, 7, 8, 11, 12]. In these systems, the haptic guidance ensures that users would not significantly deviate from the intended path. The intensity of the constrained force is considered an important factor in constraining a user's hand movement within an ideal trajectory. However, numerous other parameters can affect the user's ability to accurately trace a path. These include the force profile that maps the changes in force magnitude as a function of the distance of deviation from the path, and the effect of path or track shape on tracing performance. The contribution of this paper is largely an exploration of which parameters influence haptic path tracing. This knowledge can in turn be used by designer given the constraints of the system they are designing for. In the next section, we briefly present related work and then the results of one experiment we carried out to address the effect of these factors on haptic guidance tasks.

2 Related Work

The concept of haptic guidance can be traced back to "virtual fixture" [9]. Recently, haptic feedback for accurate path tracing has been applied to numerous applications. Tao et al. [12] demonstrated the value of a hand-writing system that teaches novices to write Chinese characters. They use haptic feedback forces that guide the hand on the right track when the student deviates from the path traces defined by the tutor.

Feygin et al. [3] evaluated the effects of different feedback conditions on users' ability to trace paths. They found that haptic+visual feedback facilitated the highest amount of learning. They also found that the shapes of the trajectories used in their study had an effect of learning performance. Similarly, Yang et al. [16] demonstrated that some trajectories of shapes are easier learned than others. Pastel [6]'s study showed that negotiating a 90° corner is more difficult than negotiating a 45° or 135° corner. Their study also revealed that users tended to make shortcut at corners by "cutting off the corner". However, it is not clear whether this is also the case when haptic guidance is provided. In particular, little is known about the extra provisions that are necessary in the haptic guidance system for assisting users to trace shapes of different curvatures.

Many of the existing haptic-guidance systems provide variable constrained force such that the force magnitude changes as a function of deviation from the path. For instance, Avizzano et al. [1] used a linear force function to generate constrained force. Kim and Yang [5] implemented various force functions (linear, logarithmic, and exponential) in their handwriting training system. But they did not conclusively suggest which types of force profiles provide sufficient accuracy for path tracing.

3 Experiment

The goal of this experiment was to evaluate the different factors that can influence the performance of path tracing with a passive haptic guidance system [16]. In particular we investigated the effect of path shape and force profile. Furthermore, it is anticipated that spring stiffness may affect tracing performance. We also confirm this in our study.

3.1 Participants and Apparatus

Nine students from a local university participated in this study. All of them were right-handed, and reported a normal to corrected-to-normal vision. None of the participants had any experience with haptic devices.

Haptic feedback was provided by a PHANToM Desktop haptic device [10]. The test system displayed the various paths that users were required to trace on a 300×300 square region in the middle of a 19 inch 1024×768 LCD display (0.4×0.4mm/pixel). Participants were asked to hold the stylus of the PHANToM device like holding a pen, and to control the experiment by pressing the button on the stylus. The shapes that users were asked to trace were two-dimensional and were laid out in the vertical plane (Fig. 1).

3.2 Procedure and Design

The study employed a 4×3×2 within-subject factorial design. The factors were path or Track Shape (TS), Force Profile (FP) and Spring Stiffness, or simply Stiffness (S). Trajectory shape was fully counterbalanced. Force profile and stiffness were presented in random order. The experiment consisted of 5 repeated trials. We collected a total of 4 (tracks) × 3 (force profiles) × 2 (stiffnesses) × 5 (trials) × 9 (participants) = 1080 trials. The participants were asked to trace a defined path with the help of haptic guidance, as accurately as possible. To ensure a complete trial, the participants were asked to indicate the start and the end of a trial by pressing the button on the stylus of the PHANToM device. Participants practiced for 5 minutes prior to the experiment to get familiar with the environment and procedures. We describe the various levels of conditions in the following sections.

Path or Track Shape. Participants were required to trace four trajectories, square, circle, S-shape, and 5-shape (Fig. 1). The tracks were generated by a graphical system and consisted of paths with a smooth curvature (circle and S-Shape) and two paths consisting of sharp corners (Square and 5-Shape).

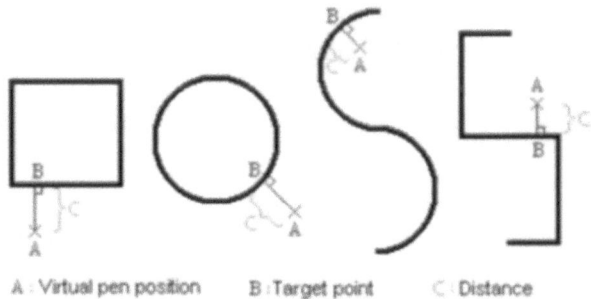

A : Virtual pen position   B : Target point   C : Distance

Fig. 1. The four paths used in the study. The S-Shape and 5-Shape paths were generated by slicing the square and circle in half and flipping the half pieces. All tracks consisted of the same length.

Force Profile. Similar to [5], we evaluated three force profile functions, a linear function, a logarithmic and a quadratic. The slope of these function is in the order that $S_{logarithmic} > S_{linear} > S_{quadratic}$ within the tolerance distance (within 1 pixel away from the ideal path), and $S_{quadratic} > S_{linear} > S_{logarithmic}$ otherwise. Force magnitude was computed using Hooke's Law, and more specifically defined with respect to the distance of deviation as follows:

$$\textbf{Force} = K_i * Function_j(Distance) \tag{1}$$

where *Distance* is the distance between end-effector position and the target position on a given trajectory (i.e. deviation from path), subscribe i is either low or high stiffness (see below), and j is either *linear*, *logarithmic*, or *quadratic* force profiles. The constants we used for each of the force profiles are provided below and resulted in the profiles presented in Fig. 2.

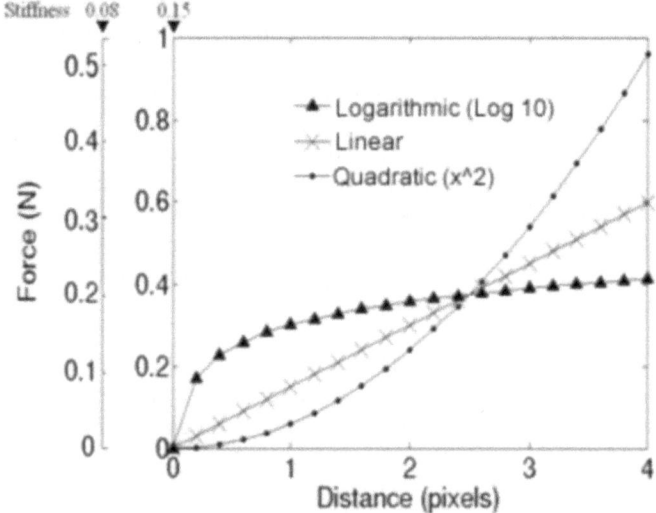

Fig. 2. Force profiles used in the experiment

Force functions which have been used are as following, where x is the distance in pixel.

$$F_{linear}(x) = ax + b, \text{ where } a=1, \text{ and } b=0; \tag{2}$$

$$F_{log}(x) = a\log_b(c*x), \text{ where } a = 1.25, b=10, \text{ and } c=40 \tag{3}$$
$$(F_{log}(x) = 0 \text{ when } x = 0);$$

$$F_{quadratic}(x) = ax^2+b, \text{ where } a=0.4, \text{ and } b=0. \tag{4}$$

Stiffness. Haptic guidance was implemented in a similar way to that described in [16]. The constrained force was generated passively so that it triggered only when the stylus deviated from the trajectory. The direction of the constraint force was always perpendicular to the direction of the hand movement. Fig. 1 depicts examples of our method for finding the target point on tracks of different shapes. The aim of this method is to guide the users' hand to the nearest point on the track. Two levels of stiffness were tested in order to confirm that spring stiffness affects users' performance on tracking given trajectories. A high stiffness or *hard spring* maintained a value of 0.15 N/mm and a low stiffness or *soft spring* consisted of a value of 0.08 N/mm.

3.3 Performance Measures

We used two metrics to assess tracking performance: Error Volume and percentage on-track. Error Volume (1) is an indicator of the average deviation distance from the path.

$$Error\ Volume = \frac{\sum_{n=1}^{N_1+N_2} Dist(X_n, X_\Gamma)}{N_2} \tag{5}$$

where X_n is the nth point on a given trajectory, N_2 is the number of points deviated form the path, and X_Γ is its corresponding point on the user trajectory. Percentage on-track (2) is a measure assessing the percentage of points that were on-track. In some applications [14], maintaining a path on the track for a large percentage of the trajectory is more important than the overall deviation amount.

$$Percentage\ on\text{-}track = (N1)/(N1 + N2) \times 100 \tag{6}$$

where N_1 is the number of points on the path. We considered a point on-track if it was within a tolerance distance (1 pixel) to its corresponding point on the ideal trajectory.

4 Results and Discussion

We used the univariate ANOVA test and Tamhane post-hoc pair-wise tests (unequal variances) for all our analyses with subjects as random factor. We present our results for both metrics Error volume and Percentage on-track.

4.1 Error Volume

There was a significant effect of Force Profile (FP) ($F_{2,\ 16} = 22.3$, $p < 0.001$), of Track Shape (TS) ($F_{3,\ 24} = 3.91$, $p = 0.02$) and of Stiffness (S) ($F_{1,\ 8} = 89.1$, $p < 0.001$) on Error volume. The error volume was larger with the soft spring and smaller with the hard spring. We found significant interaction effects between FP and S ($F_{2,16} = 17.57$, $p < 0.001$), a significant interaction between TS and S ($F_{3,\ 24} = 3.92$, $p = 0.021$) but no interaction effects between FP and TS.

Post-hoc pair-wise comparisons for FP yielded significant differences for all pairs of functions (all $p < 0.01$). Users produced fewest errors with the quadratic function than the linear or logarithmic function (Fig. 3). Post-hoc pair-wise comparisons for TS did not yield any significant differences for all pairs of functions, except between Circle and 5-Shape ($p < 0.05$). Users produced fewest errors with the S-Shape (2.867, s.e. =0.76) followed by the Circle (2.889, s.e. =0.76), the Square (3.041, s.e. =0.76) and then finally the 5-Shape (3.373, s.e. =0.76) tracks.

Our results reveal an effect of FP on error volume. A quadratic force function facilitates a lower average error volume and best deviation margins than the other two functions (see final discussion). By mapping the mean error volumes shown in Fig. 3 to the force profile functions in Fig. 2, we found that the users were able to sense the guiding force of approximately 2.5 N and to correct wrong movements by following the guidance.

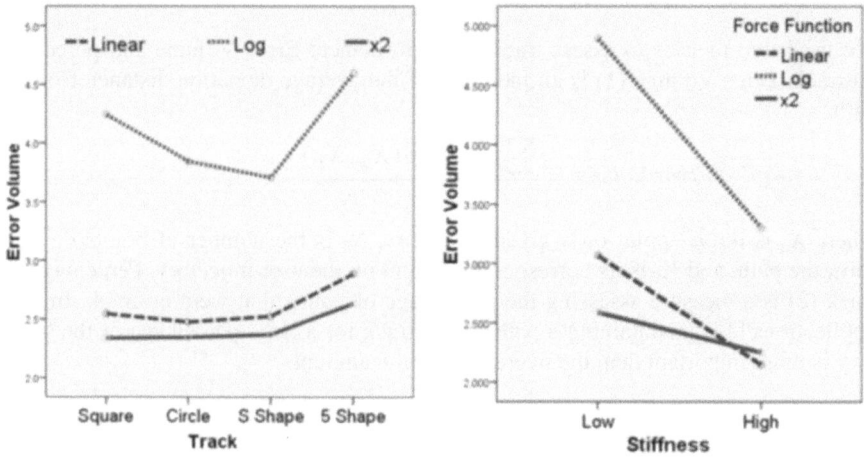

Fig. 3. Average error volume by force profile and track shape (left); and force profile and stiffness (right)

4.2 Percentage On-Track

There was a significant effect of FP ($F_{2, 16} = 174.4$, $p < 0.001$), of TS ($F_{3, 24} = 3.34$, $p = 0.036$) and of S ($F_{1, 8} = 470.5$, $p < 0.001$) on average percentage of points on-track. We found significant interaction effects between FP and S ($F_{2,16} = 32.15$, $p < 0.001$), a significant interaction between FP and TS ($F_{6,48} = 2.56$, $p = 0.031$) but no interaction effects between TS and S.

Post-hoc pair-wise comparisons for FP yielded significant differences for all pairs of functions (all $p < 0.01$). Users were able to stay on the track more frequently with the logarithmic function, then followed by the linear function and then the quadratic function (Fig 4). Post-hoc pair-wise comparisons for TS also yielded significant differences for Circle and 5-Shape ($p < 0.001$), Square and 5-Shape ($p < 0.01$) and for Circle and S-Shape ($p=0.048$). Users produced fewest deviations with the Circle (51.8%), then Square (53%), S-Shape (56.5%) and then with the 5-Shape (59%).

As observed with error volume, participants performed significantly better with the hard spring (62%) than with soft spring (49%). This confirms that the stiffness is a critical value in the design of haptically guided systems. However, it is important to be aware that the improvement of tracing performance does not grow linearly with spring stiffness (or the intensity of guiding force). We expect a leveling-off threshold, when exceeded, the improvement of tracing performance becomes less obvious.

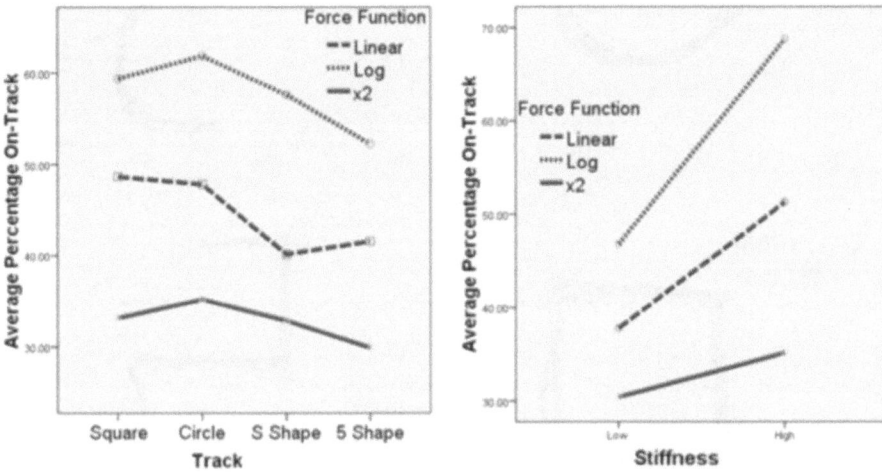

Fig. 4. Average percentage of points on track

4.3 Discussion

Interestingly, trajectory shapes with smoother curvatures revealed lower error volumes than those with sharp curves. Our observation reveals that at the edge of the curvature, users typically tend to overshoot away from the trajectory. Smooth curves do not impose this limitation as the user is making uniform movements across the trajectory. We also observe more deviation at the intersection corners in the sharper edges. This deviation is depicted in Fig. 5, showing the amount of deviation occurring at the extremities of sharp edges. Note that the finding showing that users tend to overshoot at sharp turns is different with what was previously reported in [6], in which the author revealed that users tended to make shortcut at corners. One possible explanation could be that when with haptic guidance, users tend to put less control on hand movement so that they move their hand by following the guiding force with inertia in the moving direction. Users only take over the control when significant errors occur, i.e. overshot a corner.

The path corners, can be one of the most important features of a trajectory, such as when learning a script or doing a precise maneuver. Failure to transfer corner information during motor-skill training could lead to poor outcomes. A possible way to reduce overshooting is to introduce strong constraint forces at corners to ensure the user makes the turn before going beyond the corners.

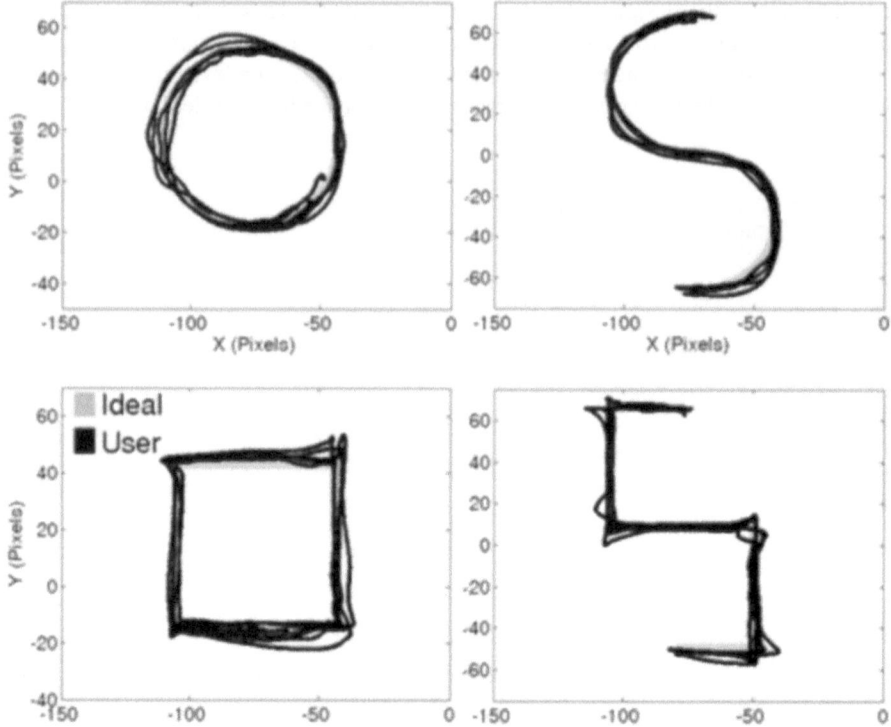

Fig. 5. Demonstration of the deviations at the intersection corners in the sharper edges

Surprisingly, while the quadratic and linear functions outperformed the logarithmic function in terms of error volume, the logarithmic outperformed both other functions with respect to percentage of points on-track. This suggests that the slope of the force profile function (see Fig. 2) has an effect on tracing performance. Furthermore, users may sense better the force when the functions have a steep slope instead of a flat slope. The linear and quadratic forces have steep slopes, so that they were more sensible to the users, and could provide better haptic constraints. As a result, the error volume for the linear and quadratic function was lower than the logarithmic function. On the other hand however, within the tolerance distance, the logarithmic function is steeper than the linear and quadratic functions. Therefore, the logarithmic generates a noticeable guiding force before the deviation becomes significant (exceeding the tolerance distance). This may explain why the user could maintain their trace on the tracks more often with the logarithmic than with the linear and the quadratic profiles.

Overall, the result suggests that designers need to take into consideration the appropriate type of force profile (and slope of the force function) based on the application in question. Ideally, the best function would be one that reduces both the error volume and facilitates the tracing of a higher number of points on track. Additionally, it is important to be aware that employing the maximum output of a force feedback device may not be necessarily helpful for achieving optimal tracing performance.

Designers need to be aware of the leveling-off threshold of the magnitude of guiding force based on different trajectories. Furthermore, designers need to employ different strategies to facilitate tracing according to the level of guidance. When provided with weak guidance, users tend to make short cut at corners. Therefore, introduction of corner threshold marks is necessary to ensure that the user crosses the corner before tracing the remainder of the trajectory. On the other hand, users tend to overshoot when following a strong guiding force. Therefore, additional constraints, such as stronger guiding force, should be useful at the corners.

5 Conclusion

In this paper we described a study measuring the effects of force profile, path shape, and spring stiffness on the performance of path tracing, in passive haptic guidance systems. The results show that all of the tested factors influence the performance of path tracing tasks. More important, the results show that users perform differently when negotiating sharp corners with or without haptic guidance. In particular, when with haptic guidance, users tend to overshot at corners. This is different to the findings reported in [6] when haptic guidance was not provided. Based on the results, we suggest that spring stiffness and/or the slope of a force profile function should be adjusted when considering changing the degree of haptic assistance at corners. High stiffness values are preferred to constrain and guide the users' movements. However, the leveling-off threshold has to be empirically determined based on the application in question. For reducing error volume and on-track percentage, a force profile with steep slope would be better suited. Overall, the linear force profile demonstrates a good balance of both error volume and on-track percentage. However, further investigation is required to identify the subtleties involved in selecting ideal force functions for tracing tasks, and the effect of these based on the underlying characteristics of the haptic guidance application.

References

1. Avizzano, C.A., Solis, J., Frisoli, A., Bergamasco, M.: Motor Learning Skill Experiments using Haptic Interface Capabilities. In: Proc. of ROMAN, pp. 198–203 (2002)
2. Crossan, A., Brewster, S.: Multimodal Trajectory Playback for Teaching Shape Information and Trajectories to Visually Impaired Computer Users. ACM Transactions on Accessible Computing 1(2) (2008)
3. Feygin, D., Keehnder, M., Tendick, F.: Haptic guidance: Experimental evaluation of a haptic training method for a perceptual motor skill. In: Proc. of HAPTICS, pp. 40–47 (2002)
4. Henmi, K., Yoshikawa, T.: Virtual lesson and its application to virtual calligraphy system. In: Proc. of ICRA, pp. 1275–1280 (1998)
5. Kim, Y.K., Yang, X.: Real-Time Performance Analysis of Hand Writing Rehabilitation Exercise in Haptic Virtual Reality. In: Proc. of Canadian Conference on Electrical and Computer Engineering, pp. 1357–1360 (2007)
6. Pastel, R.: Measuring the difficulty of steering through corners. ACM CHI, 1087–1096 (2006)

7. Plimmer, B., Crossan, A., Brewster, S.A., Blagojevic, R.: Multimodal collaborative handwriting training for visually-impaired people. In: Proc. of CHI, pp. 393–402 (2008)
8. Portillo-Rodriguez, O., Avizzano, C.A., Chavez-Aguilar, A., Raspolli, M., Marcheschi, S., Bergamasco, M.: Haptic desktop: The virtual assistant designer. In: Proc. of MESA, pp. 1–6 (2006)
9. Rosenberg, L.B.: Virtual fixtures: Perceptual tools for telerobotic manipulation. In: Proc. of the IEEE Annual Int. Symposium on Virtual Reality, pp. 76–82 (1993)
10. SensAble Technologies (1993), http://www.sensable.com
11. Saga, S., Kawakami, N., Tachi, S.: Haptic teaching using opposite force presentation. In: Proc. of the World Haptics, pp. 18–20 (2005)
12. Teo, C.L., Burdet, E., Lim, H.P.: A robotic teacher of Chinese handwriting. In: Proc. of HAPTICS, pp. 335–341 (2002)
13. Tavakoli, M., Patel, R.V., Moallem, M.: A haptic interface for computer-integrated endoscopic surgery and training. Virtual Reality 9, 160–176 (2006)
14. Webster, R., Zimmerman, D., Mohler, B., Melkonian, M., Haluck, R.: A prototype haptic suturing simulator. Medicine Meets Virtual Reality 81, 567–569 (2001)
15. Williams, R.L., Srivastava, M., Conaster, R.: Implementation and evaluation of a haptic playback system. Haptics-e 3(3), 160–176 (2004)
16. Yang, X.D., Bischof, W.F., Boulanger, P.: Validating the performance of haptic motor skill learning. In: Proc. of HAPTICS, pp. 129–135 (2008)

Haptic Interaction Techniques for Exploring Chart Data

Sabrina A. Panëels[1], Jonathan C. Roberts[2], and Peter J. Rodgers[1]

[1] Computing Laboratory, University of Kent,
CT2 7NF Canterbury, UK
{sap28,p.j.rodgers}@kent.ac.uk
[2] School of Computer Science, Bangor University,
Dean Street, Bangor Gwynedd LL57 1UT, UK
j.c.roberts@bangor.ac.uk

Abstract. Haptic data visualization is a growing research area. It conveys information using the sense of touch which can help visually impaired people or be useful when other modalities are not available. However, as haptic devices and virtual worlds exhibit many challenges, the haptic interactions developed are often simple and limited to navigation tasks, preferring other modalities to relay detailed information. One of the principal challenges of navigation with haptic devices alone, particularly single point-based force-feedback devices, is that users do not know where to explore and thus obtaining an overview is difficult. Thus, this paper presents two types of interaction technique that aim to help the user get an overview of data: 1) a haptic scatter plot, which has not been investigated to any great degree, provided by a force model and 2) a new implementation for a haptic line chart technique provided using a guided tour model.

Keywords: Haptic Visualization, Haptic Interaction Techniques, Haptic Scatter Plots, Haptic Line Charts.

1 Introduction

The use of non-visual forms of data communication, including haptics, is rapidly increasing as low cost devices become more available. Haptic technology is not only useful to increase accessibility but also provides an alternative modality to convey information (for instance when the other senses are overloaded, to provide a sensory duplication of the information or to provide different information from other modalities). Consequently, one growing application area is haptic data visualization: the use of tactile or force-feedback devices to represent and realize information and data [1]. Haptic data visualization aims to provide an alternative 'perception' or understanding of the underlying data through effective data mappings and user interactions using the sense of touch. Through the interactions with such a representation, the user can gain quantitative or qualitative understanding of some underlying data.

M.E. Altinsoy, U. Jekosch, and S. Brewster (Eds.): HAID 2009, LNCS 5763, pp. 31–40, 2009.
© Springer-Verlag Berlin Heidelberg 2009

Therefore, interaction techniques hold a key role in the interaction with and experience of the virtual world. There are several challenges of navigating haptic 3D worlds: haptics provides a lower perceptual bandwidth than vision, especially when many haptic devices are single point tools, and thus the haptic environment lacks the detail of spatial awareness that is implicit in vision. Consequently, finding new suitable metaphors and effective interaction techniques is essential, especially in the haptic visualization area, where they help the user gain an understanding of the data.

The use of haptic interactions in visualization is not widespread, as the field is young, and often involves simple interactions where haptics is limited to navigation and the auditory modality is used to convey information. This is particularly the case for charts applications. Indeed, charts are above all a visual medium as users can easily get an overview, spot trends and locate maximum and minimum values or other specific values. As the haptic modality has a much lower perception bandwidth than vision, perceiving a chart haptically is a more difficult task. For example, as most devices are single point-based, it is difficult to get an overview. Additionally, as one can lose its spatial awareness easily in a virtual 3D world, it becomes hard to make comparisons or realize values. Therefore, the challenge is to find effective representations and efficient interaction techniques that will enable understanding the underlying data.

This paper presents two new haptic interaction techniques for charts to help the user get an overview of the data: the visualization of a haptic scatter plot and a haptic line chart. Our scatter plot technique models the plot by assigning a repulsive force to each point. The user explores the plot and greater force is felt for larger concentrations of data points. For the line chart method, we implement a guidance tour, based on a museum tour metaphor first suggested by Roberts et al. [2]. Both techniques have been implemented using the force-feedback PHANTOM device.

Little research has been carried out for the haptic visualization of scatter plots. Haase and Kaczmarek [3] designed and tested the display of scatter plots both on a fingertip- and an abdomen-based eletrotactile display and conducted an experiment evaluating them. Scatter plots were created from sampling bivariate normal distributions with 25 data points and no axes were represented. Crossan et al. [4] on the other hand presented a visualization method based on haptic textures produced using granular synthesis and using the force-feedback PHANTOM device so that the user explores the textured surface that represents the plot. In contrast, our method for haptic visualization of scatter plots assigns a force depending on the distance to each point, meaning the user has a greater opportunity to explore the plot in 3D, and can detect the forces from multiple directions.

As line charts are a common representation form, their non-visual representation has previously been investigated for the exploration and extraction of features. Initially, the line was represented by a cylinder or an embossed ridge, however, users found it difficult to follow the line without slipping off the edges [5], [6]. Fritz and Barner [5] used attraction forces to keep the user on the line.

However, Yu et al. [6] proposed that engraved modelling should be used instead (also used by Van Scoy et al. [7]). Representing different lines is still an open challenge. Yu et al. [6] used different surface friction to distinguish the lines, but this misled the users at line intersections. In addition, using gridlines to convey coordinate values, as with the visual counterpart, was found to be confusing. Hence Yu et al. [8] adopted a multimodal approach, leaving haptics for navigation and dedicating the auditory modality to providing quantitative values through synthesized speech and overview through sonification. Roberts et al. [2] investigated the haptic overview and suggested (but did not implement) several different methodologies of exploration: unguided exploration (for instance a 'Friction & Zoom View' where the areas above and below the lines were assigned different textures), the constrained exploration view (i.e. the 'Exploded Data View' provides the user with different and simplified views of the data) and the guided tour. In this paper we give implementation of our version of the guidance tour.

The remainder of this paper is organized as follows. In section 2 we explain in detail our scatter plot technique. In section 3 we show an implementation of the guidance tour metaphors for the line chart in a data flow prototyping tool. Finally, in section 4 we give our conclusions and discuss further work.

2 Scatter Plots

A scatter plot is a visualization technique used to reveal the correlation between variables. By displaying data as a collection of points, the scatter plot shows the relationship between the variables through the size and location of the point cloud, the direction of the relationship, and whether outliers exist. Analyzing a scatter plot typically consists of two tasks: identifying the general trend (direction, size and position) and spotting interesting features such as outliers. This process corresponds to the visualization mantra of Ben Shneiderman namely 'Overview, zoom and filter, details on demand'. Getting an overview is therefore the first step to understanding the data. Hence we designed an overview haptic interaction technique for scatter plots that conveys the general trend of the plot.

The interaction technique was developed using a haptic interaction prototyping tool (HITPROTO) based on the haptic scene-graph H3D API [9]. The tool uses visual configurable blocks to represent the flow of data. There are two main blocks categories: the flow blocks (Wait For, Everytime, Switch) which are either event-based (e.g. "Wait For the device's button to be pressed") or testing conditions ("If the key pressed is equal to 1") and the actions blocks (adding/removing effects, creation/modification of haptic and guidance effects). Setting the parameters of these blocks and linking them together enables to produce an executable scenario. X3D[1] scenes can also be 'loaded' to use a scene representation or act on it (for example by highlighting shapes or removing them from the scene).

---

[1] X3D [10] is an ISO open standard scene-graph design and the successor to the VRML standard. It is used and extended within H3D to specify the 3D scene-graph of the world and particularly the geometric and basic haptic properties.

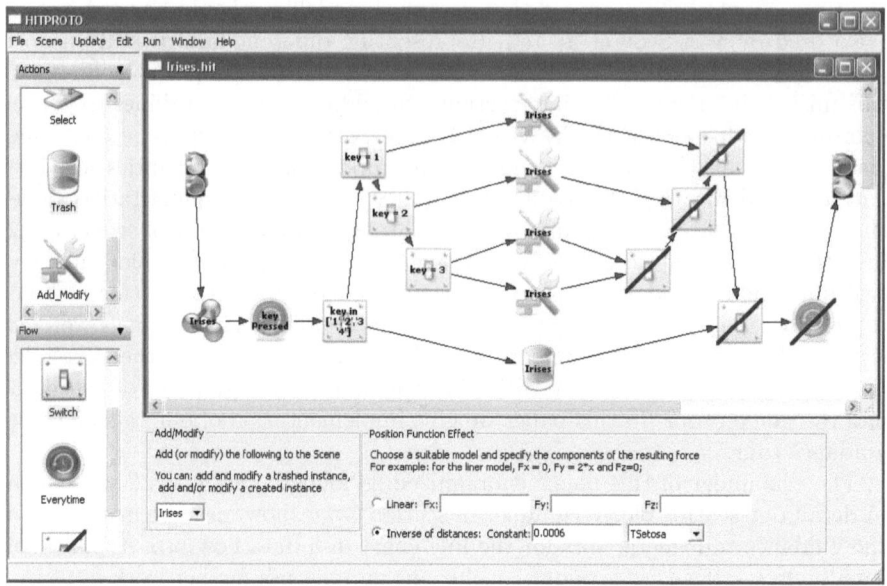

Fig. 1. Diagram showing the data flow of the scatter plot visualization of Irises. The haptic effect is first created, then when a key is pressed, the haptic effect is either set for a particular grouping node (for all of the Irises or each of them) or removed depending which key was pressed.

Figure 1 shows the data flow diagram of the scenario for the haptic visualization of a scatter plot. The data used was the Iris dataset by R.A. Fisher [11]. Three species of Iris flower (Setosa, Versicolor and Virginica) have been studied and particularly for the following parameters: sepal width, sepal length, petal width and petal length.[2] Both 2D and 3D charts were generated to highlight the correlation of the flowers sepal length and petal length/width (see Figure 2). The scenario's concept is simple, each dataset is associated with a key on the keyboard ('1','2','3') and pressing that key 'adds' the haptic effect to the corresponding dataset (see Figure 1). The user can then feel the datasets successively as well as the whole dataset. We believe that in doing so the user will get a general idea about the location of the different datasets relative to each other as well as their respective size as it provides simplified and different views of the data [2].

The force model used is part of the PositionFunctionEffect block in the HIT-PROTO tool where the user can either specify the 3D components of the force or use a predefined force model, which can be applied to grouping nodes in the scene graph, according to the device's position. For the scatter plot scenario, the predefined model was used (see Figure 3 and 4), which computes the resultant repulsive force to be applied as the sum of the inverse of the distances from the

---

[2] The Iris dataset was obtained from the XmdvTool website, in the data sets category[12].

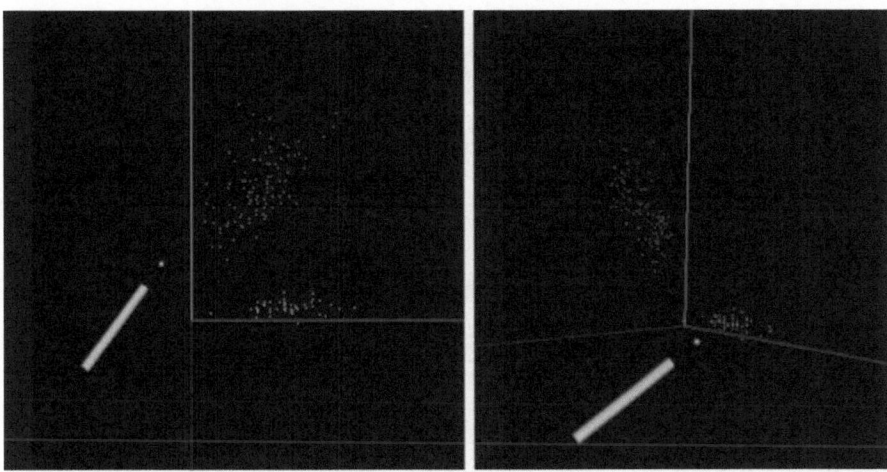

Fig. 2. Screenshot showing the 3D visual display of the Iris datasets used for the haptic visualization as well as a rotated view

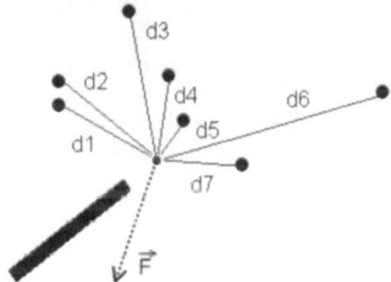

Fig. 3. Diagram describing the force model used in the scatter plot visualization scenario. The resultant force is equal to: $\boldsymbol{F} = -k \times \sum_{i=1}^{7} \frac{1}{d_i} \times \boldsymbol{u_i}$ with k a constant and d_i being the distances to the points from the device and $\boldsymbol{u_i}$ the unit vectors from the device to the points.

haptic device to each point from the point cloud in the chosen grouping node along with the sum of the unit vectors between the device and these points (see Equation 1 and Figure 3 for an example).

$$\boldsymbol{F} = -k \times \sum_{i=1}^{n} \frac{1}{d_i} \times \boldsymbol{u_i} \ . \tag{1}$$

with d_i, the distance from point i to the device and $\boldsymbol{u_i}$, the unit vector of the vector from the device to point i.

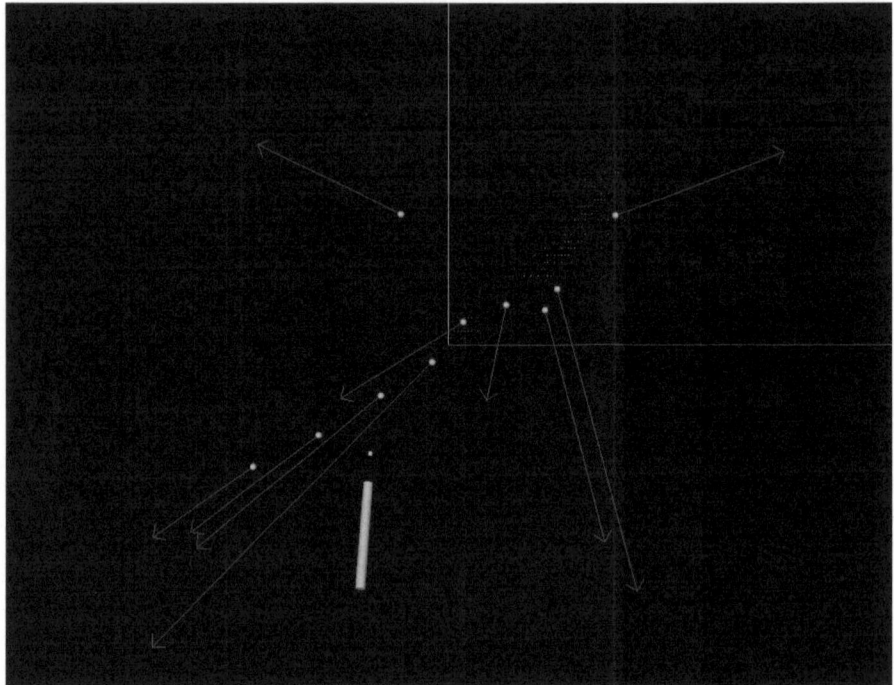

Fig. 4. To illustrate the forces on the device, this figure gives a 2D representation of the force model using a 2D Iris dataset. The larger filled circles show the device positions and the vectors show the force applied on the device. The magnitude of the force was scaled down for the figure.

The initial idea to convey the scatter plot's general trend was a function that depends on the number of points within a given radius. However, simply using the number of points as a factor led to jerky forces thus highlighting the need for a continuous force that would give a smoother feeling. After experimenting and discarding different models (including one trying to use the device's velocity vector), we decided to use all the points of the dataset, and not only the ones present in the device's tip vicinity, along with the continuous model based on the inverse of the distances from each point to the device. The idea was that the closer the device would get to a highly point-concentrated area, the greater the force factor would be, while being further away from the points reduces the force (see Figure 4). This model can convey relatively well the direction of the point cloud and relative position to the other datasets (when felt successively as in the scenario presented above and in Figure 1). However, it is often difficult to get an accurate picture of the point cloud density.

The axes are not represented haptically and thus cannot be used directly with the scatter plot representation as the user can get lost easily in 3D space. Nonetheless, the axes could be included using a guidance interaction technique, as detailed in the following section, by taking the user to the centre of the axes and then to the first point(s) of the point cloud to help the user locate the point cloud within the world coordinate system.

3   Line Charts

Line charts are one of the most common representations for statistical data. However, many challenges still remain for their exploration with non-visual techniques. Researchers resort to solving these using the auditory modality (as explained in section 1), however this modality may not be appropriate or available, and so a pure haptic technique would be preferable as an alternative. The challenges are similar to that for the scatter plot: getting an overview of the chart and locating and comparing specific features (such as minimum and maximum values, intersections). Getting an overview haptically is difficult, especially when using point-based devices.

We believe that guidance coupled with free exploration can contribute to building a better mental image of the chart. To that effect, three different guidance tours have been developed (initially presented as a poster at the HAID'07 workshop [13]) using the HITPROTO tool. In this paper, we describe an alternative implementation of the museum tour metaphor, where the user is taken along a predefined path

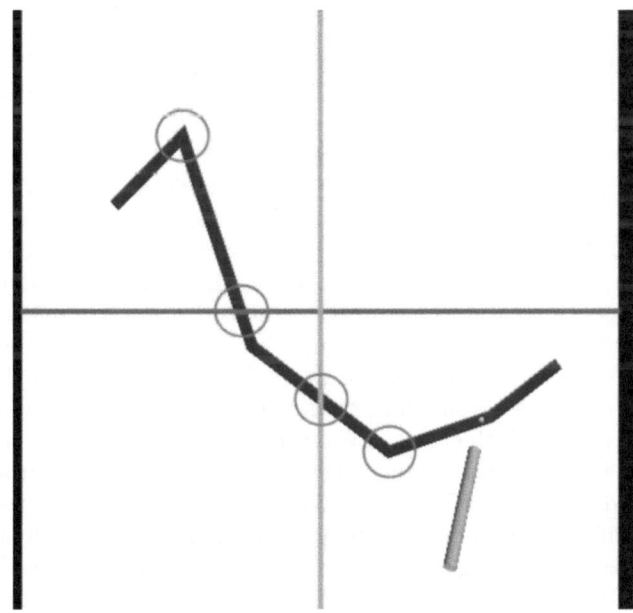

Fig. 5. The 3D visual display of the line chart, presenting a V-shaped line and embossed axes on a chart surrounded by walls. The device is led along the shown engraved line, but stops for a given time to allow for user exploration in the areas indicated by circles, which are max, min points and intersection with axes.

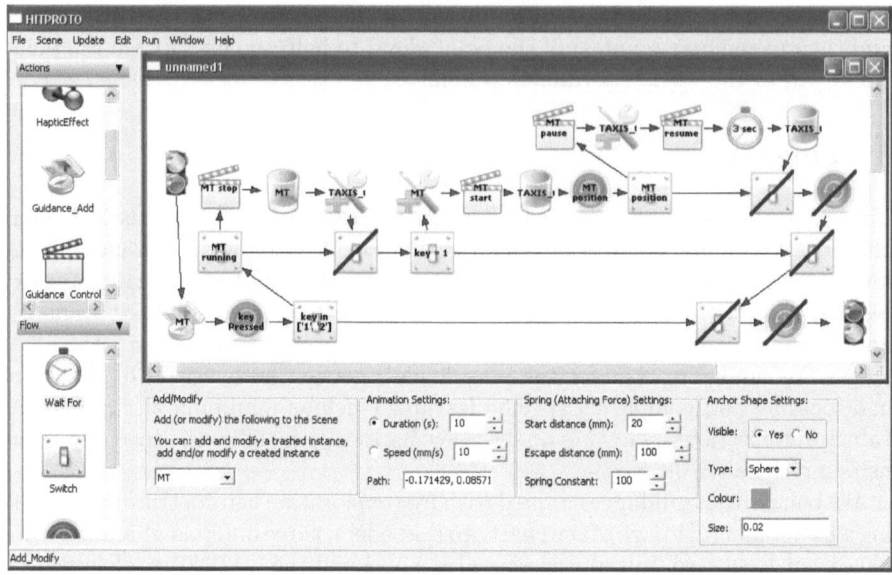

Fig. 6. Diagram showing the data flow of the museum tour scenario in the case of a line chart visualization. The tour stops at all the points of interests: maximum, minimum and intersection with the axes.

and stops at chosen points of interest. The user can then roam around to explore the surroundings for a given amount of time before the tour resumes. The tour relies on a spring force to keep the user attached to a moving 'anchor'.

Unlike previous work ([6], [8]), the representation chosen occupies the whole chart display (not only positive values) and in addition to the V-shaped line, some 'bumps' are added at the intersections of the V-shape line with the axes (see Figure 5). Figure 6 shows the data flow diagram of the museum tour scenario. The museum tour interaction is the combination of a standard guidance block and the behaviour at points of interests during the movement. The guidance block enables setting general parameters, such as the parameters for the attaching force, the path and speed or cycle time (see Figure 6 as these general parameters are repeated in the 'Add_Modify' block). The behaviour at the points of interest is tuned with the block 'Everytime', set on monitoring the movement positions, the block 'Switch' testing whether the user is passing through a point of interest and the blocks 'Guidance_Control' to specify the actions of pausing (with the roaming around) and resuming the guidance. Additionally in this case, the axes are removed during the guidance and added back whenever the guidance is interrupted (by pausing it or by triggering the free exploration mode).

4   Conclusion

This paper has presented two types of interaction technique that aim to help the user get an overview of data. The first technique displays scatter plot data and uses a force model based on the sum of the inverse of the distances to each point from the current user location to feel the data. The interaction technique scenario described allows feeling the different datasets successively, thus providing the user with the relative positions of the datasets and their general direction. The second technique describes guidance tour interactions, in particular a museum tour metaphor, in the context of line chart data. The user is taken along the V-shaped line and can explore the surroundings at points of interest, such as minima/maxima and intersections with the axes, where the tour pauses for a given time. Different variations of the guidance interactions (waterskier tour, bus tour, museum tour) can be used to provide different types of information (the whole line overview, the points of interests, the surroundings).

Future work might consist in perfecting the force model for the scatter plot interaction technique, as the ones presented lack accuracy for the point density and possibly finding and testing new models such as one using the inverses of the squared distances as commonly used in physics. Equivalent models could be found for tabular data. As for the guidance interaction techniques, more behaviour at the points of interest will be investigated in the case of line charts. Examples of such behaviour ideas include adding magnetic lines leading to the axes to provide an estimate of the coordinate values as well as possibly allowing the user to leave haptic landmarks to find the points easily during free exploration. The guidance techniques could also be extended to other types of charts, such as bar charts, where the path could take the user along each bar and back to the x-axis or along the envelope of the bar chart. The two interaction techniques described should also be evaluated. More generally, other types of interaction techniques should be researched to provide chart overview and also to provide quantitative and qualitative values.

References

1. Roberts, J.C., Panëels, S.A.: Where are we with Haptic Visualization? In: World Haptics, 2nd Joint EuroHaptics Conference and Symposium on Haptic Interfaces for Virtual Environment and Teleoperator Systems, pp. 316–323. IEEE Press, Washington (2007)
2. Roberts, J.C., Franklin, K., Cullinane, J.: Virtual Haptic Exploratory Visualization of Line Graphs and Charts. In: Bolas, M.T. (ed.) The Engineering Reality of Virtual Reality. Electronic Imaging Conference, vol. 4660B, pp. 10–19. IS&T/SPIE (2002)
3. Haase, S.J., Kaczmarek, K.A.: Electrotactile Perception of Scatterplots on the Fingertips and Abdomen. Med. Biol. Eng. Comput. 43, 283–289 (2005)
4. Crossan, A., Williamson, J., Murray-Smith, R.: Haptic Granular Synthesis: Targeting, Visualisation and Texturing. In: 8th International Conference on Information Visualisation, pp. 527–532. IEEE Press, Washington (2004)
5. Fritz, J.P., Barner, K.E.: Design of a Haptic Graphing System. In: The Rehabilitation Engineering and Assistive Technology Society of North America Conference. RESNA (1996)
6. Yu, W., Ramloll, R., Brewster, S.A.: Haptic Graphs for Blind Computer Users. In: Brewster, S., Murray-Smith, R. (eds.) Haptic HCI 2000. LNCS, vol. 2058, p. 41. Springer, Heidelberg (2001)

7. Van Scoy, F.L., Kawai, T., Darrah, M., Rash, C.: Haptic Display of Mathematical Functions for Teaching Mathematics to Students with Vision Disabilities: Design and Proof of Concept. In: Brewster, S., Murray-Smith, R. (eds.) Haptic HCI 2000. LNCS, vol. 2058, pp. 31–40. Springer, Heidelberg (2001)
8. Yu, W., Cheung, K., Brewster, S.A.: Automatic Online Haptic Graph Construction. In: Wall, S.A., Riedel, B., Crossan, A., McGee, M.R. (eds.) EuroHaptics, pp. 128–133 (2002)
9. The H3D API, http://www.h3dapi.org/
10. The X3D Standard, http://www.web3d.org/
11. The Data and Story Library (DASL),
    http://lib.stat.cmu.edu/DASL/Stories/Fisher'sIrises.html
12. XmdvTool Datasets, http://davis.wpi.edu/~xmdv/datasets/iris.html
13. Panëels, S., Roberts, J.C.: Haptic Guided Visualization of Line Graphs: Pilot Study. In: Workshop on Haptic and Audio Design, poster session (2007)

Audio Bubbles: Employing Non-speech Audio to Support Tourist Wayfinding

David McGookin, Stephen Brewster, and Pablo Priego

Department of Computing Science
University of Glasgow, Glasgow G12 8QQ
{mcgookdk,stephen}@dcs.gla.ac.uk
www.dcs.gla.ac.uk/~mcgookdk

Abstract. We introduce the concept of Audio Bubbles - virtual spheres filled with audio that are geocentered on physical landmarks, providing navigational homing information for a user to more easily locate the landmark. We argue that the way in which tourists navigate is not well supported by traditional visual maps, and that Audio Bubbles better support the serendipitous discovery and homing behaviours exhibited in such tourist activities. We present a study comparing Audio Bubbles to a visual map in a real world navigation task. Navigation with Audio Bubbles appeared to be faster and was preferred by most of the participants. We discuss the findings and outline our future development plans.

Keywords: Non-visual Navigation, Wayfinding, Auditory Display.

1 Introduction

The evolution of hand-held mobile computers into generic, powerful, location-aware connected devices, has dramatically increased their utility as mobile navigation and tourism guides. This has allowed dynamic contextual information to be easily provided to the user. However, most of these applications, whilst allowing dynamic overlay of tourist related information, such as restaurants, and other points of interest, present this information on a visual display. Most commonly this is achieved by overlaying the information onto a visual map. This requires the user to look at the map to locate, and then navigate towards, points of interest. The user is required to make an explicit decision to find a feature on the map and then determine how to navigate towards it. Whilst this is useful when the user has a pre-determined goal in mind, there are many situations when being a tourist where there are no concrete destination goals.

Brown and Chalmers [1] carried out an ethnographic study of tourist wayfinding and navigation in Edinburgh in the UK. Visiting activity was found to be planned, but only up to a point. They found that the user would "wander" along the streets and serendipitously discover interesting sights and sounds. In picking a restaurant for example, tourists would select an area with a number of restaurants and wander about until they found one that met their needs. The user would not find a particular restaurant in a guidebook and then constantly refer to a map

M.E. Altinsoy, U. Jekosch, and S. Brewster (Eds.): HAID 2009, LNCS 5763, pp. 41–50, 2009.
© Springer-Verlag Berlin Heidelberg 2009

to head towards that restaurant. Similar requirements, that mobile guides should not be over controlling or constraining, were found by Kassinen [2] and Cheverst *et al.* [3]. However, Brown and Laurier [4] identified situations where a tourist would study a map out of context in an effort to learn an unfamiliar area and its points of interest. An obvious conclusion to draw is that the user does not want to be explicitly burdened with constant referral to a visual map, but at the same time does not want to miss important (to the tourist) points of interest in an area that is being explored. This is something that existing visual maps fail to adequately support, primarily as the user must make an explicit decision to refer to the map, which is unlikely in such a "wandering" scenario.

Another situation where traditional visual maps fail to properly support wayfinding is when the user must relate a position of a point of interest in a printed map or guidebook, to his or her current position in the environment [1]. This context change is most common at the very end of a route. For example, identifying which building on a street was the birthplace of a famous author, when the user is standing on the same street. At this point, traditional street level maps provide poor support for the user to "home in" on the desired point of interest. Brown and Chalmers [1] describe an example of finding a particular building in amongst other buildings, but many other points of interest may be equally unobtrusive, and considerably smaller in the environment (e.g. the zero kilometer point in Place du Parvis in Paris - represented by a small brass disc on the ground). Again, visual maps are not particularly suitable to support this activity.

Both of these situations, wandering and homing, are related - the user may not have an explicit goal in mind, preferring to find interesting features as he or she walks. However, when the user identifies an interesting feature or landmark, perhaps from a street sign, a handed paper flyer, or the pre-study of guidebooks [4], it can be difficult to use a visual map to home in on the point of interest. In both cases the visual nature of maps (both paper and electronic) means that they are not the most appropriate way to support these types of activity, either due to context changes, or that when wandering there would be little need for constant referral to maps.

2   Audio Bubbles

To better support the kinds of activities previously mentioned, we have developed a technique called Audio Bubbles. These provide serendipitous, non-visual awareness and "homing" information onto nearby points of interest, without the explicit user intervention that would be required with a visual map. An Audio Bubble is a virtual sphere geolocated on a real world point of interest. Within the bubble audio information is played. This audio is manipulated to indicate the user's distance from the point of interest that the bubble is centred upon. As a user walks around the environment, he or she will come into contact with the boundary of the Audio Bubble, and become aware of a nearby point of interest, or that a pre-determined point of interest is being approached. Using

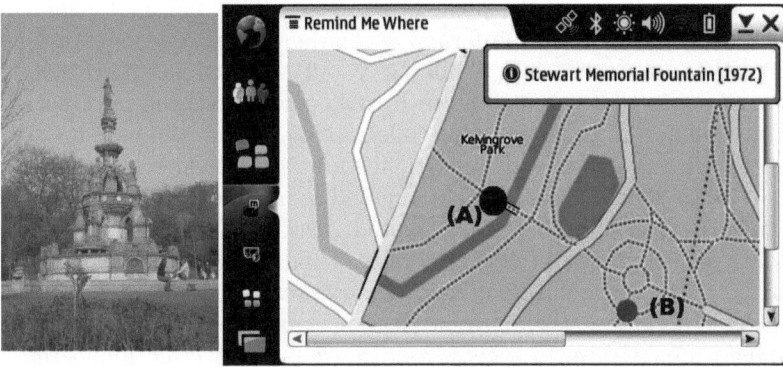

Fig. 1. Right: A screenshot of the visual display showing the current user location (A) and the location (B) and name of the next landmark). **Left:** A picture of the landmark.

the distance information encoded in the audio, the user is guided the last few meters onto the point of interest. In this way, the Audio Bubble acts as a form of augmented reality, overlaying virtual audio in the soundscape of the environment, and increasing awareness of nearby interesting things that can either be attended to or ignored, without the need to make the explicit decision to look for points of interest that is required with a visual map.

We explicitly make two decisions with Audio Bubbles and our future implementations. Firstly, that the sound we use within the bubble is unspecified. There are several types of sound that could be used to fill an Audio Bubble, however what sound is used will depend on the use to which the bubble is being put, and the point of interest it represents. In some cases a sound representative of the point of interest would be most appropriate, but as with many other areas of auditory display, such sounds may not be possible and a different kind of sound may need to be chosen. We discuss this issue later in the paper. Another decision with the design of Audio Bubbles is that they should augment the existing soundscape in an environment. Both Brown and Chalmers [1] and De Certeau [5] point out that the sights, sounds and smells - the ambiance of an area - is one of the main aspects of a location a tourist wishes to experience.

3   Related Work

Whilst there has been a great deal of work on visual navigation (e.g. [3] [4]), less has been done on auditory navigation. The bulk of this work on auditory navigation has investigated the presentation of route and nearby environmental information to people with visual impairments (e.g. [6]). One of the first systems for sighted users was Holland *et al's.* [7] AudioGPS system. This used spatial panning, with the user wearing headphones to communicate accurate direction of a specified point of interest. To communicate distance, a Geiger counter metaphor was used, with a sound repeating at an increasing rate as the user became closer to the point of interest. Whilst informal comments on AudioGPS proved positive, no formal evaluation was carried out. Strachan *et al's.* [8] gpsTunes worked in a similar way, but modified the perceived spatial location of music that the user listened to to provide bearing information and modified volume to provide distance information. Again, the user had to wear headphones to support the spatialisation. An informal field evaluation was carried out which showed promise that the audio manipulation may be useful in navigation. Reid *et al's. Riot! 1831* [9] study used similar techniques to present an auditory soundscape of the Bristol riots of 1831 in the streets of modern day Bristol. As users walked around a city square wearing headphones and a GPS equipped mobile device, their current locations were used to trigger sound effects and vignettes of real world events from the riots. Reid *et al.* found that their system provided a deep level of immersion within the experience. Stahl's [10] Roaring Navigator used a similar technique, with spatialised sounds of animals being used to indicate the directions of, and guide visitors to, various enclosures in a zoo. Software running on HP iPaq mobile computers was used to determine the user's position and to adjust the sounds so that they appeared to be coming from the direction of the appropriate enclosure. Participants could either browse, hearing the nearest four enclosures, or could navigate by hearing only the destination sound source. Again, Stahl's system required the user to wear headphones, but shared the audio being played so that users could interact with each other and use the sound as a shared reference point. Stahl notes that in audio museum guides, visitors often do not interact with each other due to the wearing of headphones causing social isolation. An evaluation of the system was performed with encouraging user feedback, however Stahl did not compare performance with a standard visual map of the zoo.

Whilst these previous systems show that there are uses of auditory wayfinding, they also have issues when being applied as solutions to the issues of tourist navigation discussed in the introduction. Firstly, they all use spatialised audio, which works by varying the waveforms that reach the left and right ears, allowing the user to determine the direction of the sound source. However, such spatialisation requires the user to wear headphones. Headphones can have the effect of blocking out the existing auditory soundscape. In some cases this may be desirable, such as on a crowded commuter train; in other cases it may be less desirable, such as wandering through the streets of an historic Italian town. It is very much a preference of the user whether to wear headphones or not. Additionally, as noted by Stahl, wearing headphones can cause the wearer to be more socially isolated from his or her fellow travellers. Given that travel is a social activity [1], this social isolation would not be desirable. Whilst there are other types of headphone such as bone conductance [11], which do not occlude the ears and other sounds reaching them, they are not common or widely available. As we would like Audio Bubbles to augment the existing environment (and certainly not block it) and that some users may not wish to wear headphones, we consider that monaural presentation (without headphones) via the loudspeakers on the mobile device presents a "worst case" scenario, and worthy of investigation.

Secondly, whilst users can perform navigation using the audio, none of the studies discussed have compared performance between visual only and auditory navigation. Does the addition of sound lead to superior performance in near field navigation over navigation with a visual map? Finally, the sounds used in these systems always present some sort of audio no matter how far away the user is from the point of interest. As already described, this is unlike Audio Bubbles, which have a defined outer exterior beyond which no sound is heard. This is important as if the user has no defined route, such as when wandering, the audio that is presented must be controlled in some way so that too much is not presented concurrently, and the overall effect is not cacophonous [12].

4   Evaluation

In our initial evaluation we chose to investigate only the "homing" scenario as discussed in the introduction. Primarily, this is because we wanted to gain feedback on the size and utility of the Audio Bubbles. Once we have investigated this scenario we can consider more fully the utility of the Audio Bubbles and investigate the wandering scenario, where the user has no fixed destination in mind. Given the discussion of the previous work, we sought to answer two related research questions about Audio Bubbles:

RQ1: Can Audio Bubbles presented using monaural (without headphones) sound be effective navigation aids to help users locate points of interest?

RQ2: Do Audio Bubbles improve navigation performance over a visual only map?

Our initial Audio Bubbles implementation was developed as an application on a Nokia N810 Internet Tablet. We used the built-in GPS system on the tablet to determine the position of the user. The visual display of the device showed a scrollable overview map of the area (obtained from www.openstreetmap.org), the current user position and one or more landmarks (see Fig.1). The Audio Bubbles were implemented along the Geiger counter principle of Holland *et al.* [7]. Distance was mapped to the repetition rate and volume of a short "click" sound. The closer the user was to the point of interest, the louder and more frequently the click was played. We set the initial boundary to be 250 meters from the point of interest, and no sound was played out with this. To evaluate our initial Audio Bubble implementation, we carried out a within groups study on a route navigation task. Eight participants (4 women and 4 men) aged 24-27 took part in the experiment. All of the participants reported normal hearing and normal, or corrected to normal sight. None were paid for participation. Each participant walked a route using a visual map representation (Visual Condition), or the same map representation but with each landmark encased in an Audio Bubble (Audio Bubble Condition). The order in which participants carried out each condition was counterbalanced. Two different routes (of length 940m and 810m respectively) around different landmarks (4 and 3 respectively) in a local park were used. Routes were counterbalanced between conditions to avoid

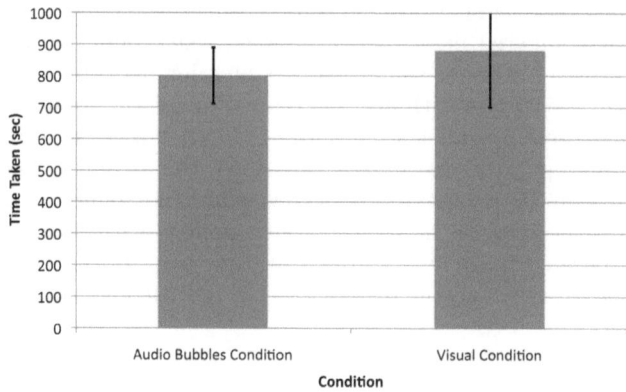

Fig. 2. Graph showing the mean time taken in both the Visual and Audio Bubbles Conditions. Shown with one standard deviation.

difference effects. We chose the park as it has a number of small points of interest, such as statues, that are easy to miss even when close by. These are the sorts of points of interest that we consider the Audio Bubbles to be particularly useful to find, as they are not obvious in the environment. This is similar to the situations where a user is trying to home onto a very old pub, or a niche museum in a city centre, but keeps walking past as these are unobtrusive landmarks in the environment. None of the participants were regular visitors to the park. Participants were shown how to use the system for the current condition and told that they should navigate to the landmarks given. The position of the first landmark and its name was displayed on the map, as was the participant's current location (see Fig. 1). In the Audio Bubbles condition only the Audio Bubble for the current point of interest was active. As the participant entered a zone approx 10m around the landmark, a trumpet sound was played, signalling that the landmark had been correctly identified. The next landmark was then shown on the display. When all landmarks had been found, the participant was given a message that the experiment was over. At the end of the experiment the participant was verbally debriefed and asked for any comments.

4.1 Results

As participants were required to reach each point of interest, accuracy was uniform. In addition, we recorded the time taken to complete each route. One participant's results were found to be over two standard deviations from the mean for each of the conditions. We judge this participant as an outlier and exclude those results from the analysis. The mean time taken for the other seven participants is summarised in Fig. 2. As the groups were not fully counterbalanced we elected not to perform a formal statistical analysis of the time taken. However the results do seem to indicate that there was some influence of the Audio Bubbles in reducing the time taken to find the points of interest.

4.2 Qualitative Results

The interviews with participants after the experiment yielded valuable insights into the usefulness of the Audio Bubbles, and how they compared to the Visual condition. Participants were asked a number of open-ended questions about their experience with both of the systems. These included the usefulness of the Audio Bubbles, as well as how the sounds changed as the participant moved around the environment. The participants were also asked for any further issues that they felt relevant.

Five of the eight participants stated that they found the Audio Bubbles helpful in trying to locate the points of interest. One participant (the participant whose results were excluded from those in Section 4.1) did not find the Audio Bubbles helpful, and two participants were unsure if the Audio Bubbles were helpful or not. All of the participants were able to determine the changes in sound as they walked through the Audio Bubbles. Comments primarily addressed the "range" of the Audio Bubbles, that is how far away from a point of interest the user should be before entering the bubble. Two of the participants found that the sound started too far away from the point of interest and was initially annoying, but with decreasing distance from the point of interest the sound became more useful and valuable. One participant saying *"I would use it, yeah. ... But the beeping would be annoying if it would beep for every little sight. I would turn off the sound until I got closer to the points "*, the same participant went on to say that the sound cues were useful *"especially when you were closer, when you were closer it really gave you hints I think"*. In the experiment a user entered an Audio Bubble when around 250 meters from the point of interest. However, in this situation where the user already has a destination determined, the Audio Bubble would only be useful (as discussed in the introduction) when close to that destination. Whereas, in a wandering scenario, where the Audio Bubble notifies the user of a nearby point of interest that he or she may not be aware of, a larger radius may be more appropriate. Further investigation is needed to provide concrete guidance in this area.

In conclusion, we can consider to have provided some answer to RQ1, that Audio Bubbles were a useful way to navigate and locate points of interest. However, we cannot make claim to have answered RQ2, that Audio Bubbles are superior to simple visual maps. The results we obtained appear to show that the Audio Bubbles allowed some reduction in time taken for the fine grained homing behaviour participants needed at the end of a route, and that the Audio Bubbles were certainly not poorer than only using a visual map. However, the results are not definitive and further investigation, given the qualitative feedback from users, is required to refine Audio Bubbles to be truly useful navigational aids.

5   Future Work

From the experimental results there are several refinements that should be carried out on the Audio Bubbles, and several future avenues for investigation.

5.1   Audio Bubble Size

Many of the participants stated that they found the audio started to play too early. In other words the Audio Bubble was too big. As sound can be annoying, determining the appropriate size of the Audio Bubble, so that it is neither too large or too small is important. In our initial implementation we chose 250m as the boundary of the bubble, primarily down to informal testing by the experimenter. While reducing the size of the bubble would be beneficial based on the comments by participants, we also need to consider the role the Audio Bubble plays at a particular time, and if its size should be adjusted accordingly. In our experimental situation, the Audio Bubbles were centred on locations that the participant was trying to reach and were felt to be useful in "homing" onto the point of interest when nearby. However when wandering, the purpose of the Audio Bubble is primarily to inform the user of a nearby, non-obvious point of interest, that he or she might be interested in. In such a situation, an Audio Bubble with a radius extending to a "reasonable walking distance" between the user and the point of interest, may be more appropriate. Further research into how the size of the Audio Bubble should change according to user context is therefore required.

5.2   Investigating Different Types of Audio for Points of Interest

In our experiment we chose to follow the same auditory design as that of Holland *et al.* [7], using a Geiger counter metaphor, where audio pulses repeat at a faster rate the closer the user is to the point of interest. We also increased the volume level of the pulses as the user came closer to the point of interest. We chose to investigate this kind of mapping as it provided more information in the sound as to how far the user was from the point of interest. As we did not use spatial presentation (see Section 3), we wanted to ensure that the distance information was communicated as clearly as possible. In addition, unlike the work of Stahl [10], many of our points of interest had no obvious sound that could be used to represent them, meaning a more abstract sound was required. The issue being that although the sound communicates distance information, it contains no information as to the point of interest being navigated towards. In our initial study this is not so important, but it would be useful to communicate some information about the point of interest in the sound, thereby assisting the user in the decision to walk towards that point of interest or not. In practice, it is likely that there will be a mixture of points of interest which can have appropriate natural sounds associated with them, and points of interest for which there is no obvious sound. Therefore, it would be necessary to use different auditory representations to represent different points of interest, and further investigation is required to identify how well these different auditory representations work with our monaural system.

5.3   In-Field Evaluations

Our evaluation work described in this paper has sought to investigate both the utility of Audio Bubbles as an idea, and consider how they can be used in the homing scenario discussed in the introduction. However, another goal of our work is to allow discovery of points of interest in a wandering scenario - where the user has no particular goal or destination in mind, and would tend not to use a visual map as a means of navigation. Our intention is that Audio Bubbles would allow him or her to "stumble" upon interesting places that may not be obvious in the built environment. However, our current implementation is not advanced enough to be used as a full tourist application. Therefore, we are building an Apple iPhone application so we can evaluate the wandering scenario with real tourists.

5.4   Social Points of Interest

Our experiment, as well as all of the existing work in Section 3, assumes that points of interest are fixed and determined by some prior authority. However with the growth of the mobile Web and geotagging (associating a geographic location with data) of information on the Web, we can add and augment points of interest from other users of the system. Axup and Viller [13] have looked at how backpacking is a temporally shared activity amongst travellers, and Goel [14] has looked at how information on tourist guides can be shaped and filtered by social networks. In addition, websites such as www.tripadvisor.com allow users to rate and comment on their experiences of hotels and restaurants. How should this information be represented through Audio Bubbles? One approach would be to modify the size of the Audio Bubble based on recommendations. For example, a well recommended restaurant would have a larger Audio Bubble than one with a poorer recommendation. However this may, given the results of our study, be annoying, as the audio would begin to play too far away from the point of interest. An alternate approach would be to use speech as the sound in the Audio Bubble, reading out the review comments that previous visitors have left. Which is better is unclear, and further investigation is needed in this area to identify the best approach.

6   Conclusions

In this paper we have introduced the concept of Audio Bubbles - virtual spheres filled with audio that are geocentered on physical points of interest to provide serendipitous discovery and assist in fine grained navigation. We argue such a technique is both necessary and useful, as visual maps fail to adequately support the non-directed wandering and homing activities that tourists engage in. The outcome of our experiment does indicate that the use of the Audio Bubbles provided some advantage to users when trying to locate points of interest in a park. Future versions of the Audio Bubbles will seek to identify how flexible the concept is, by evaluation in less directed navigational scenarios than those in our experiment, and as such will provide answers to many of the issues raised in our future work section. In conclusion, we believe that with further refinement to the auditory design, future versions of the Audio Bubbles will provide significant advantages to users engaged in loosely directed navigation, in a way that is useful, but not intrusive or annoying, and support the undirected wandering behaviours that tourists currently engage in.

Acknowledgements. This work is supported by EU FP7 No.224675 "Haptimap".

References

1. Brown, B., Chalmers, M.: Tourism and mobile technology. In: Eighth European Conference on CSCW, Helsinki, Finland, vol. 1, pp. 335–354. Kluwer Academic, Dordrecht (2003)
2. Kaasinen, E.: User needs for location-aware mobile services. Personal and Ubiquitous Computing 7, 70–73 (2003)
3. Cheverst, K., Davies, N., Mitchell, K., Friday, A., Efstratiou, C.: Developing a context-aware electronic tourist guide: some issues and experiences. In: CHI 2000, The Hague, The Netherlands, vol. 1, pp. 17–24. ACM, New York (2000)
4. Brown, B., Laurier, E.: Designing Electronic Maps: An Ethnographic Approach. Mapbased Mobile Services - Theories, Methods and Implementations, vol. 1. Springer, Heidelberg (2005)
5. De Certeau, M.: The Practice of Everyday Life. University of California Press (1984)
6. Wilson, J., Walker, B.N., Lindsay, J., Cambias, C., Dellaert, F.: Swan: System for wearable audio navigation. In: ISWC 2007, Boston, MA, vol. 1 (2007)
7. Holland, S., Morse, D.R., Gedenryd, H.: Audiogps: Spatial audio in a minimal attention interface. Personal and Ubiquitous Computing 6(4), 6 (2001)
8. Strachan, S., Eslambolchilar, P., Murray-Smith, R.: gpstunes - controlling navigation via audio feedback. In: MobileHCI 2005, Salzburg, Austria, vol. 1, pp. 275–278. ACM, New York (2005)
9. Reid, J., Geelhoed, E., Hull, R., Carter, K., Clayton, B.: Parallel worlds: Immersion in location-based experiences. In: CHI 2005, Portland, Oregon, vol. 2, pp. 1733–1736. ACM Press, New York (2005)
10. Stahl, C.: The roaring navigator: A group guide for the zoo with shared auditory landmark display. In: MobileHCI 2007, Singapore, vol. 1, pp. 282–386. ACM, New York (2007)
11. Marentakis, G.: Deictic Spatial Audio Target Acquisition in the Frontal Horizontal Plane. PhD thesis, University of Glasgow (2006)
12. McGookin, D.K., Brewster, S.A.: Dolphin: The design and initial evaluation of multimodal focus and context. In: ICAD 2002, Kyoto, Japan, vol. 1, pp. 181–186 (2002)
13. Axup, J., Viller, S.: Augmenting travelgossip: Design for mobile communities. In: OZCHI 2005, Canberra, Australia, vol. 2. ACM, New York (2005)
14. Goel, A.: Urban pilot: A handheld city guide that maps personal and collective experiences through social networks. In: Tanabe, M., van den Besselaar, P., Ishida, T. (eds.) Digital Cities 2001. LNCS, vol. 2362, pp. 384–397. Springer, Heidelberg (2002)

Interactive Sonification of Curve Shape and Curvature Data

Simon Shelley[1], Miguel Alonso[1], Jacqueline Hollowood[2], Michael Pettitt[2], Sarah Sharples[2], Dik Hermes[1], and Armin Kohlrausch[1]

[1] Human Technology Interaction, Eindhoven University of Technology, P.O. Box 513, NL 5600 MB, Eindhoven, The Netherlands
`d.j.hermes@tue.nl`
[2] Human Factors Research Group, The University of Nottingham, University Park, Nottingham, NG7 2RD UK

Abstract. This paper presents a number of different sonification approaches that aim to communicate geometrical data, specifically curve shape and curvature information, of virtual 3-D objects. The system described here is part of a multi-modal augmented reality environment in which users interact with virtual models through the modalities vision, hearing and touch. An experiment designed to assess the performance of the sonification strategies is described and the key findings are presented and discussed.

Keywords: sonification, sound synthesis, modal synthesis, virtual environments, haptics, human-computer interaction.

1 Introduction

The work presented in this paper describes how sound is used in a multimodal design environment in the context of the European project called: Sound And Tangible Interfaces for Novel product design (SATIN). The aim of the project is to develop a multimodal interface consisting of an augmented reality environment in which a product designer is able to see *virtual* 3-D objects and to both explore and modify the shape of these objects by directly touching them with his/her hands. This visual–haptic interface is supplemented by the use of sound as a means to convey information and feedback about the virtual object and the user interaction. More specifically, the use of sonification allows the designer to explore geometric properties related to the surface of the object that are hardly detectable by touch or sight.

The sonification of the surface *shape* and *curvature* of the virtual object is a specific requirement of the SATIN project. Shape and curvature data are of special interest to designers when considering the aesthetic quality of the object's surface. In industrial design, for example, a smooth surface that is continuous in terms of curvature is highly desirable because discontinuities in curvature result in a disconnected appearance in the light reflected from the surface. Designers

M.E. Altinsoy, U. Jekosch, and S. Brewster (Eds.): HAID 2009, LNCS 5763, pp. 51–60, 2009.
© Springer-Verlag Berlin Heidelberg 2009

refer to surfaces with continuous curvature as *Class A surfaces* and they consider them as being aesthetically appealing [1].

This article is organized as follows. Section 2 of this paper presents a short summary of work related to this study. Then in section 3 we propose and describe the sonification strategies. This is followed by section 4 where the evaluation and results are presented. Finally, a summary of the work and a discussion of the results are given in section 5.

2   Related Work

Previous work involved physical models that are able to synthesize sounds that are perceptually associated with interactions between humans and physical objects [2,3,4]. In studies involving interaction with virtual objects this type of auditory feedback can be used to convey important environmental information in order to improve perception and performance [5,6].

One of the requirements of the SATIN project is the interactive sonification of numerical data related to the geometrical properties of a virtual object. Another related study suggests the use of sonification as a means to present geometrical data of surfaces in scientific and engineering applications [7]. However, so far in this context we are not aware of any study carrying out a formal evaluation of the success of such an approach.

In the current literature there exists a number of studies with a similar aim, but in a variety of different contexts. For example, some attempts have been made to sonify numerical data for people with visual impairment; in some cases this is combined and/or compared with kinaesthetic methods used to represent the same data [8,9,10,11]. Perhaps one of the earliest studies was made by Mansur who used a sinusoidal wave, the frequency of which is used to sonify the ordinate value of a numerical function where the abscissa is mapped to time. Mansur discovered that mathematical concepts such as symmetry, monotonicity and the slopes of lines could be determined using sound after a relatively small amount of training [10]. Another related application domain involves the sonification of financial data [12,13,14]. The aim of such systems is to inform the user about financial information, for example stock market prices.

Recently, researchers have also been interested in a more general approach to sonification, in which the aim is to produce sonification toolkits. Such toolkits are designed to allow users to map arbitrary data to sound with relative ease. Examples of sonification toolkits include *Listen* [15] and *Muse* [16], developed by Lodha, *Musart* [17], developed by Joseph and Lodha and the *Sonification Sandbox* [18], developed by Walker and Cothran. However, these toolkits are not suitable for the SATIN project because, in their current state, they rely on pre-defined data sets. In the case of the SATIN project the data are produced dynamically by the interaction of the user with other elements of the system.

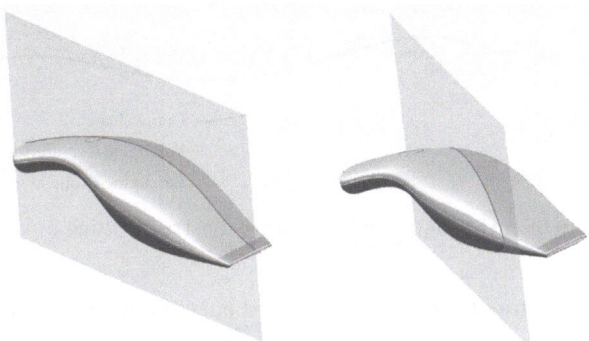

Fig. 1. Example of a virtual 3-D object with two potential 2-D cross-sectional slices highlighted

3   Sonification of Geometrical Data

In the SATIN system, although dealing with 3-D models, the haptic interaction is limited to a 2-D cross-sectional slice of the virtual object's surface. The user is free to select this slice from any part of the 3-D object. An example of two different cross-sections of the same 3-D virtual object is illustrated in Fig. 1. In this work we consider two specific requirements of the SATIN project, which are the sonification of data relating to the *curve shape* and *curvature* of the surface of the virtual object along the cross-sectional plane of interest. It is important to note that curve shape and curvature information are not sonified simultaneously, but prior to the operation of the system the user can select which of these two geometrical properties will be sonified. The aim of the sonification system is to inform the user about the chosen geometrical property of interest at the point of contact between the user's finger and the haptic device. The system is designed so that only one point of contact along the 2-D cross section can be evaluated using sound at any one time. The data to be sonified are described below.

- **Curve shape:** We consider only a 2-D cross-sectional slice of the virtual object's surface such as those highlighted in Fig. 1. As a result, the curve shape is a plane curve C represented by a two-dimensional function. For example, an arbitrarily generated curve shape is shown in Fig. 2, which illustrates the shape of an object's surface along the cutting plane.
- **Curvature:** For a given plane curve C, the curvature at point p has a magnitude equal to the reciprocal of the radius of the osculating circle, i.e. the circle that shares a common tangent to the curve at the point of contact. The curvature is a vector that points to the centre of the osculating circle. For a plane curve given explicitly as $C = f(p)$ the curvature is given by:

$$\mathcal{K} = \frac{\frac{d^2 C}{dp^2}}{\left(1 + \left[\frac{dC}{dp}\right]^2\right)^{\frac{3}{2}}}. \tag{1}$$

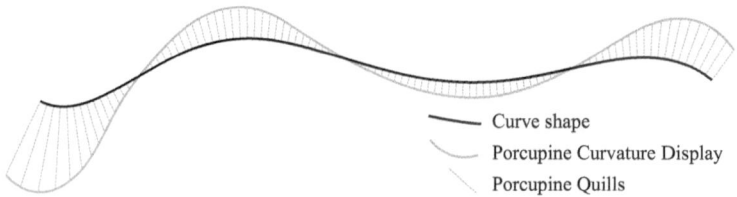

Fig. 2. Illustration of curve shape with curvature illustrated using a *porcupine* display

In Fig. 2 the curvature of the afore mentioned curve shape is illustrated by using a technique referred to as *porcupine* display. Porcupine quills are perpendicular to the curve and their lengths are proportional to the curvature at the point of contact with the curve shape, as given by (1). Therefore the longer the porcupine quill line, the higher the value of curvature, i.e. the tighter the curve. Note that in this example if the porcupine quill is drawn on top of the curve shape this denotes a negative curvature value, i.e. a convex shape, whereas a porcupine drawn underneath the curve shape denotes a positive curvature value, i.e. a concave shape.

Modes of interaction
During the sonification of the curve shape and curvature of a virtual object, three different modes of haptic interaction are considered, as illustrated in Fig. 3. The first involves the impact between the user's finger and the model at a certain point, for example by tapping the object, as shown in Fig. 3(a). The second, Fig. 3(b), involves a sustained contact with the object without movement. Finally, the third type of interaction, Fig. 3(c), involves sliding a finger over the object.

Fig. 3. Modes of user interaction with the virtual model

3.1   Sonification Approaches

The selection of sound synthesis approaches and mapping strategies to sonify the geometrical data of interest may have a significant effect on how the information is perceived by the user. The users expressed a wish for the sound that is produced to be influenced by the mode of interaction in an intuitive way. In an attempt to address this, we developed a post-processing module which we refer to in this article as the *kinetic module*. This module post-processes the generated sound according to the interaction between the user's fingers and the haptic device, i.e., the speed of the finger and pressure it exerts on the device. Furthermore, three different sonification strategies were designed. These strategies and the kinetic module are described later in the text.

As the final haptic interface is in development, we are currently using a digital drawing tablet as an alternative input device. Contact between the drawing tablet pen and the tablet is analogous to contact between the user's finger and the haptic device in the SATIN project. The position of contact along the x-axis of the drawing tablet is equivalent to the position of finger contact along the length of the selected cross-sectional profile represented by the haptic interface. In order to develop and prototype the sonification platform, we use the Max/MSP graphical development environment.

3.2   Mapping Strategy

In each of the three sonification methods, the value of the geometrical parameter of interest, curve shape or curvature, is mapped to the fundamental frequency of the carrier sound. The mapping is implemented in such a way that the minimum absolute value of the geometrical parameter is mapped to a minimum frequency of 200 Hz and the maximum absolute value found in the dataset is mapped to a maximum frequency of 800 Hz. Additionally, linear changes in the geometrical parameter of interest are mapped to logarithmic changes in the sound carrier frequency. After experimenting with a number of mappings, users considered this to be the easiest to understand. In fact, frequency is a common choice for this type of application and it is considered to be particularly strong as a means to sonify changing values [19]. Humans perceive frequency changes with a relatively high resolution. Typically, frequency changes of pure tones can be perceived with an accuracy of up to 0.3% [20]. The following sub-sections describe the three sonification strategies used in this study, named after the sound synthesis technique used to sonify or *carry* the information of interest.

3.3   Sinusoidal Carrier

For this approach, a pure sinusoidal tone is used as the carrier and its frequency is modulated according to the mapping strategy described in the previous subsections.

3.4   Wavetable Sampling Carrier

For this strategy, a continuous sound is produced whilst the user remains in contact with the haptic device. This sound is made up of a harmonically rich recording of a bowed cello. The attack and decay parts of the note are removed from the sound excerpt, leaving only the middle part with a stable fundamental frequency. The fundamental frequency of the recorded note is 131 Hz, which closely corresponds to C3 on the Western musical scale. This sound is continuously played back in a loop, in such a way that no audible artefacts or interruptions are present.

Given the harmonically rich nature of the cello sound, the technique referred to as Time-Domain Pitch-Synchronous Overlap-Add (TD-PSOLA) provides a high quality yet computationally efficient way to modify the fundamental frequency of the sound [21]. Using this technique, the fundamental frequency of the cello sound is mapped to the geometrical property of interest according to the mapping strategy defined in sub-section 3.2.

3.5   Physical Modelling Carrier

In this case the carrier sound is generated using a *modal synthesis* approach. Modal synthesis is a physical modelling technique for sound rendering, with theoretical roots in modal analysis [22]. The aim of modal synthesis is to mimic the dynamic properties of an elastic structure in terms of its characteristic *modes* of vibration. The modal synthesis implementation employed for this work is based on research by Van den Doel and Pai [23]. A more detailed explanation of this approach and its implementation in the SATIN project is given in [24].

For this study, a model of a circular plate was implemented, resulting in a sound with a rich but in-harmonic spectrum. The modal frequencies of the plate were calculated by solving the wave equation for a circular plate [25]. In order to sonify the geometrical data we map the magnitude of the parameter of interest to the fundamental frequency of the modal synthesis model, according to the mapping strategy described in sub-section 3.2. The perceptual effect of frequency scaling in this way gives the impression that, as the fundamental frequency increases, the modelled object decreases in size.

3.6   Kinetic Module

The kinetic module is an optional addition to the sonification system that post-processes the generated sound according to the interaction between the user's fingers and the haptic device. When the kinetic module is switched on, the intensity and the spectral tilt of the selected sound are varied according to the pressure and exploration speed exerted by the user's finger. In addition, the sound is passed through a high-pass filter with a cut-off frequency that is varied according to the pressure of the user's finger. This technique is based on a perceptual study, the details of which can be found in [26]. Finally, with the kinetic module switched on, if the user taps the haptic interface, as illustrated by Fig. 3(a), a short impulse of the selected sound is produced. Note that, apart from varying its intensity, the kinetic module has very little perceptual effect on the sinusoidal carrier, which consists only of a single spectral component.

4   Evaluation and Results

This paper presents a selection of results from two separate studies, performed as part of a wider experiment aimed at investigating the use of sound in the SATIN system [27]. These studies were designed to evaluate the performance of the sonification strategies as a means to communicate both curve shape and curvature. Twelve test subjects participated in the evaluation, with the requirement that they all have a basic understanding of calculus. In both studies all three of the sound carriers described above were used and compared. In the case of

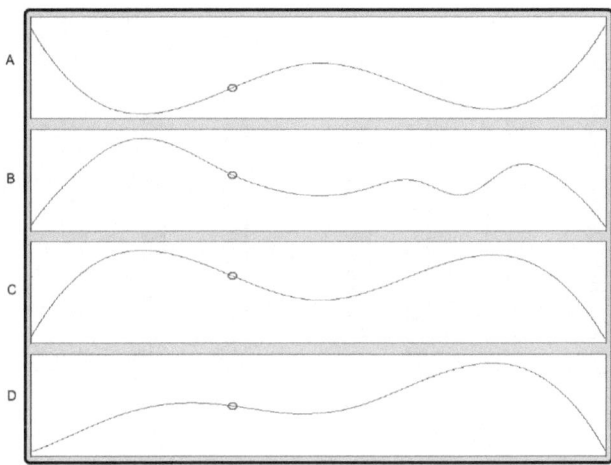

Fig. 4. Screen-shot showing an example of the multiple choice options presented to the test subjects. The small circle indicates the position on the curve that the user is exploring at that time.

wavetable sampling carrier and the physical modelling carrier we also tested the effect of applying the kinetic module. For the first experiment a number of different synthetic curve shapes were generated. The subjects were invited to explore the curve shape using a stylus on a digital drawing tablet as input, as described earlier. At the same time they were presented with the graphical representations of four different curve shapes, of which only one corresponds correctly to the sonified data. From these options, they were then asked to choose which of the curves best corresponds to the sounds that resulted from their exploration using the stylus. Fig. 4 is a screen-shot showing an example of the multiple choice options presented to the test subjects. The circle is used to convey to the test subject the position on the curve that he/she is exploring at that time.

For the second experiment, the same group of subjects were invited to another session, in which they were again presented with the graphical representations of four curve shapes. This time the data sonified corresponds to the curvature of one of the presented curve shapes. As an example, for the curve illustrated in Fig. 2, the subject would see the black curve but would listen to the sonification of the grey curve. In this case the subjects were given a short explanation of how curvature is related to curve shape.

Fig. 5 shows the average percentage of correct answers and response times for each sonification strategy and for both experiments. A significant finding is that each of these strategies, both with and without the kinetic module, performed at an equivalent level of about 80% accuracy. In addition, as a result from questionnaires administered during the tests, it was discovered that the subjects exhibited a preference for the sinusoidal and wavetable sampling carrier approaches. It was also found that participants needed in the region of 30 seconds to make their decision in each of the multiple choice questions.

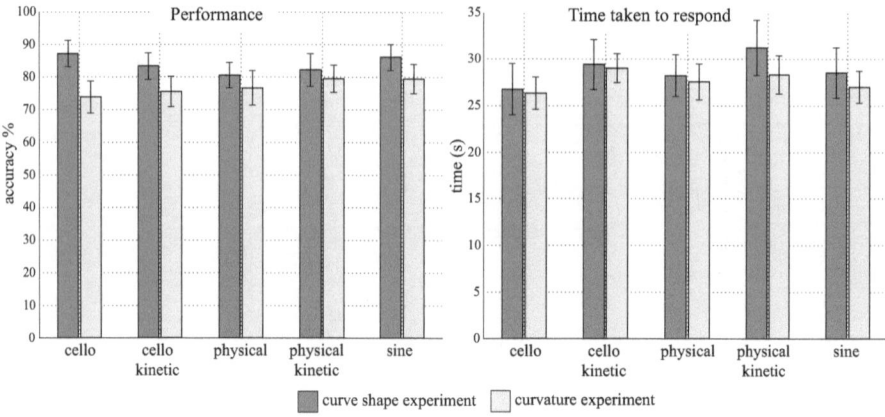

Fig. 5. Judgement accuracy and response time for both curve shape and curvature experiments

5   Conclusions

In this paper we present a sonification approach to communicate information about the curve shape and curvature of a virtual 3-D object. The described auditory system is part of a multi-modal augmented reality environment where designers can interact with 3-D models through the additional modalities of vision and touch. We implemented three sonification approaches that share a common mapping scheme, in which the parameter to sonify is mapped to the fundamental frequency, or the set of modal frequencies, of the sound carrier. The three approaches consisted of sinusoidal, wavetable sampling and physical modelling sound carriers. In addition, a novel concept is presented in which the sound that is produced is intuitively influenced by the way that the user explores the virtual object. A post-processing module, referred to in this article as the kinetic module, has been developed to implement this concept. A number of experiments have been conducted in order to evaluate the performance of these sonification strategies in communicating both curve shape and curvature information. The results, for a set of twelve participants, show a consistently high level of performance in all cases of about 80% accuracy for both curvature and curve shape. Participants required on average around 30 seconds to choose their answer in the multiple choice test. Finally, in this specific implementation and context, we could not find any advantage of using the kinetic module either in terms of performance or user preference.

The main contribution of this work is to show the feasibility of communicating geometrical data through sonification. We have attempted to validate this concept by means of a formal evaluation. We consider the obtained results to be highly encouraging since they suggest that humans, previously unexposed to the system, are capable of establishing a clear connection between the sound and the underlying data. This relatively high level of performance is consistent for the curvature experiment. We consider this to be a remarkable result, as in this case the sound is related to one of the presented images through the non-trivial mathematical expression given by (1), rather than the direct relationship exhibited in the curve shape experiment.

Acknowledgments

This research work has been partially supported by the European Commission under the project FP6-IST-5-054525 SATIN – Sound And Tangible Interfaces for Novel product design and by the COST Action IC0601 on Sonic Interaction Design. The authors also wish to thank Patrick Bosinco for providing Fig. 1.

References

1. Catalano, C.A., Falcidieno, B., Giannini, F., Monti, M.: A survey of computer-aided modeling tools for aesthetic design. Journal of Computing and Information Science in Engineering 2, 11–20 (2002)
2. Rocchesso, D., Fontana, F. (eds.): The Sounding Object. Mondo Estremo (2003)
3. Van den Doel, K., Kry, P.G., Pai, D.K.: Foleyautomatic: Physically-based sound effects for interactive simulation and animation. In: Proceedings of ACM SIGGRAPH, Los Angeles, CA, USA, August 2001, pp. 12–17 (2001)
4. Van den Doel, K., Pai, D.K.: The sounds of physical shapes. Presence 7(4), 382–395 (1998)
5. Díaz, I., Hernantes, J., Mansa, I., Lozano, A., Borro, D., Gil, J., Sánchez, E.: Influence of multisensory feedback on haptic accessibility tasks. Virtual Reality 10, 31–40 (2006); Special Issue on Multisensory Interaction in Virtual Environments
6. Kjœr, H.P., Taylor, C.C., Serafin, S.: Influence of interactive auditory feedback on the haptic perception of virtual objects. In: Proceedings of the 2nd International Workshop on Interactive Sonification, New York, UK (2007)
7. Minghim, R., Forrest, A.R.: An illustrated analysis of sonification for scientific visualisation. In: Proceedings of the 6th IEEE Visualization Conference, pp. 110–117 (1995)
8. Kennel, A.R.: Audiograf: a diagram-reader for the blind. In: Proceedings of the second annual ACM conference on Assistive technologies, pp. 51–56 (1996)
9. Stevens, R.D., Edwards, A.D.N., Harling, P.A.: Access to mathematics for visually disabled students through multimodal interaction. Human–Computer Interaction 12(1), 47–92 (1997)
10. Mansur, D.L., Blattner, M.M., Joy, K.I.: Sound graphs: A numerical data analysis method for the blind. Journal of Medical Systems 9, 163–174 (1985)
11. Roth, P., Kamel, H., Petrucci, L., Pun, T.: A comparison of three nonvisual methods for presenting scientific graphs. Journal of Visual Impairment & Blindness 96(6), 420–428 (2002)
12. Mezrich, J., Frysinger, S., Slivjanovski, R.: Dynamic representation of multivariate time series data. Journal of the American Statistical Association 79(385), 34–40 (1984)
13. Janata, P., Childs, E.: Marketbuzz: Sonification of real-time financial data. In: Proceedings of the 10th International Conference on Auditory Display, Sydney, Australia, July 6-9 (2004)

14. Mauney, B.S., Walker, B.N.: Creating functional and livable soundscapes for peripheral monitoring of dynamic data. In: Proceedings of the 10th International Conference on Auditory Display, Sydney, Australia, July 6-9 (2004)
15. Wilson, C.M., Lodha, S.K.: Listen: A data sonification toolkit. In: Proceedings of the 3rd International Confonference on Auditory Display, Palo Alto, USA (1996)
16. Lodha, S.K., Beahan, J., Heppe, T., Joseph, A.J., Zne-Ulman, B.: Muse: A musical data sonificaton toolkit. In: Proceedings of the 4th International Confonference on Auditory Display, Palo Alto, USA (1997)
17. Joseph, A.J., Lodha, S.K.: Musart: Musical audio transfer function real-time toolkit. In: Proceedings of the 8th International Conference on Auditory Display, Kyoto, Japan (2002)
18. Walker, B.N., Cothran, J.T.: Sonification sandbox: A graphical toolkit for auditory graphs. In: Proceedings of the 9th International Conference on Auditory Display, Boston, MA, USA (2003)
19. Stockman, T., Nickerson, L.V., Hind, G.: Auditory graphs: A summary of current experience and towards a research agenda. In: Proceedings of the 11th International Conference on Auditory Display, Limerick, Ireland (July 2005)
20. Wier, C.C., Jesteadt, W., Green, D.M.: Frequency discrimination as a function of frequency and sensation level. Journal of the Acoustical Society of America 61, 178–183 (1977)
21. Moulines, E., Charpentier, F.: Pitch-synchronous waveform processing techniques for text-to-speech synthesis using diphones. Speech Communication 9(5/6), 453–467 (1990)
22. Rossing, T.D., Fletcher, N.H.: Principles of Vibration and Sound, 2nd edn. Springer, Heidelberg (2004)
23. Van den Doel, K., Pai, D.K.: Modal Synthesis for Vibrating Objects. In: Audio anecdotes: tools, tips, and techniques for digital audio. AK Press (2004)
24. Alonso, M., Shelley, S., Hermes, D., Kohlrausch, A.: Evaluating geometrical properties of virtual shapes using interactive sonification. In: IEEE International Workshop on Haptic Audio visual Environments and Games, pp. 154–159 (2008)
25. Leissa, A.W.: Vibration of Plates, NASA SP-160. NASA, Washington (1969)
26. Hermes, D.J., van der Pol, K., Vankan, A., Boeren, F., Kuip, O.: Perception of rubbing sounds. In: Proceedings of the International Workshop on Haptic and Audio Interaction Design, Jyväskylä, Finland (2008)
27. Hollowood, J., Pettitt, M., Shelley, S.B., Alonso, M.A., Sharples, S., Hermes, D.: Results of investigation into use of sound for curvature and curve shape perception. Technical report, internal deliverable to SATIN project FP6-IST-5-054525 (2009)

Accessing Audiotactile Images with HFVE Silooet

David Dewhurst

daviddewhurst@hfve.org
www.HFVE.org

Abstract. In this paper, recent developments of the HFVE vision-substitution system are described; and the initial results of a trial of the "Silooet" software are reported. The system uses audiotactile methods to present features of visual images to blind people. Included are details of presenting objects found in prepared media and live images; object-related layouts and moving effects (including symbolic paths); and minor enhancements that make the system more practical to use. Initial results are reported from a pilot study that tests the system with untrained users.

Keywords: Vision-substitution, sensory-substitution, HFVE, Silooet, blindness, deafblindness, audiotactile, haptics, braille, Morse code.

1 Introduction and Background

HFVE Silooet software allows blind people to access features of visual images using low-cost equipment. This paper will first summarise the system, then report on the latest developments, and finally describe initial results of a pilot study that tests the system.

At the 1st HAID Workshop, the HFVE (Heard & Felt Vision Effects - pronounced "HiFiVE") vision-substitution system was shown exhibiting areas of images via speech and tactile methods, with demonstration shapes also shown [1]. At the 3rd HAID Workshop the "Silooet" (Sensory Image Layout and Object Outline Effect Translator) software implementation was shown presenting predetermined object outlines and corners of items present in a sequence of images [2]. Recent developments include:- presenting found or predetermined objects; symbolic moving effects and layouts; and minor enhancements such as an adapted joystick, and methods for rapidly creating and presenting simple images and diagrams in audiotactile format. A pilot study of the system has recently commenced.

The HFVE project is not focused on a specific application, but is trying various methods for presenting sequences of visual images via touch and sound. The main approach used differs from other methods which, for example, allow people to explore a shape or image by moving around it under their own control. Instead, the HFVE system generally "conducts" a user around an image, under the control of the system (albeit with user-controlled parameters), which might be less tiring and require less attention of the user than when requiring them to actively explore an image. (The system could be used in combination with other approaches.)

M.E. Altinsoy, U. Jekosch, and S. Brewster (Eds.): HAID 2009, LNCS 5763, pp. 61–70, 2009.
© Springer-Verlag Berlin Heidelberg 2009

Other work in the field includes tone-sound scanning methods that have been devised for presenting text [4], and for general images [5]; and software for presenting audiotactile descriptions of pixels in computer images [6]. Audio description is used to supplement television, theatrical performances etc. (The merits of other approaches are not discussed in this paper.)

2 System Features

The HFVE system aims to simulate the way that sighted people perceive visual features, and the approach is illustrated in Fig. 1:-

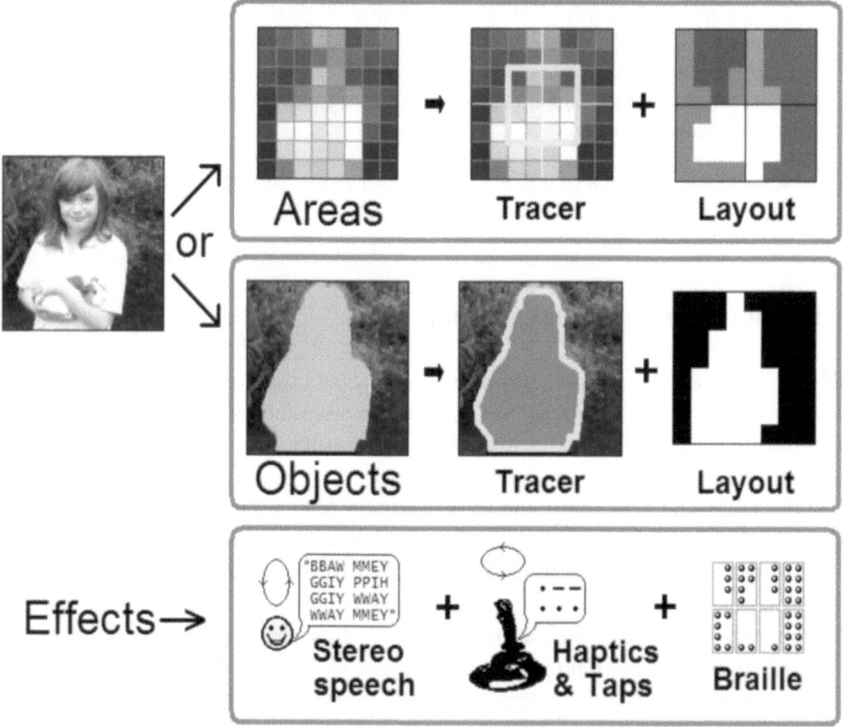

Fig. 1. Presenting features of a visual image as Area or Object "tracers" and "layouts"

For any image, or section of an image, the property content (colour, texture etc.) of "Areas" can be presented; or the properties of identified "Objects". (The term "object" is used to refer to a specific entity that is being presented, for example a person, a person's face, part of a diagram, a found coloured "blob" etc., whether found by the system, or highlighted in material prepared by a sighted designer.) For both "Areas" and "Objects", the information is presented via moving audiotactile effects referred to as "Tracers" - for Areas, the tracer path shows the location of the areas, and for Objects the path shows the shape, size, location and (if known) the identity of the objects. "Layouts" present the arrangement of (usually two) properties within an Area or Object, and normally use a regular grid-like format Fig. 1.

The paths of the tracers are presented:- via apparently-moving sounds, that are positioned in "sound space" according to location, and pitched according to height; and via a moving force-feedback device that moves/pulls the user's hand and arm – in both modalities the path describes the shape, size and location (and possibly identity) of the Areas or Objects. As the system outputs both audio and tactile effects, users can choose which modality to use; or both modalities can be used simultaneously.

The properties (colours, textures, types etc.) of the Areas or Objects are either presented within the audiotactile tracers, or separately. In the audio modality, speech-like sounds generally present categorical properties (e.g. "boo-wuy" or "b-uy" for "blue and white"). In the tactile modality, Morse-code like "taps" can be presented on a force-feedback device, or alternatively a separate braille display can be used Fig. 1. The "layout" of properties is best presented on a braille display, though, as will be described later, there are practical ways of presenting certain object layouts via speech or taps. (Appropriate mappings for speech etc. have previously been reported [1,2,3]). Until recently, layouts were used for presenting the arrangements of properties in rectangular Areas. However the content of objects can be also presented via layouts.

A key feature of the system is the presenting of corners/vertices within shapes, which initial tests show to be very importing in conveying the shape of an object. Corners are highlighted via audiotactile effects that are included at appropriate points in the shape-conveying tracers.

Although one possible tracer path for presenting an object's shape is the object's outline Fig. 1, other paths such as medial lines and frames can be used Fig. 5. "Symbolic Object Paths" are found to be effective, as they present the location, size, orientation and type of object via a single tracer path.

3   Recent Developments

3.1   Presenting Predetermined and Found Objects

HFVE Silooet can present both objects found in images "on the fly", and predetermined objects from prepared media. Fig. 2 illustrates the process:- for non-prepared media (e.g. "live" images) the system attempts to Find (a) objects according to the user's requirements, and builds a "Guide" (b) of the found objects. Alternatively a previously-prepared "Guide" (b) can be used to give the objects and features that are present. Finally, the corresponding Effects (c) are presented to the user.

The system can use predetermined Guide information if available, otherwise it attempts to find appropriate objects, and if none are found, it outputs area layouts.

Fig. 2. The HFVE system processing approach

For prepared media, a sighted designer can highlight the entities present in images, and identify what they are, their importance etc. Such predetermined entity information can be embedded in files of common multimedia formats (e.g. MP3 or AVI). The combined files are produced via a straightforward procedure, and they can also be played on standard media players (without audiotactile effects).

The predetermined sequences are presented as a series of one or more "Views" that comprise a scene, the set of Views being referred to as a "Guide". For each View, one or more objects can be defined. These can be marked-up on bitmap images, each bitmap containing groups of non-overlapping objects. Usually one bitmap will contain the background objects, and one or more further bitmaps will handle the foreground and details Fig. 3.

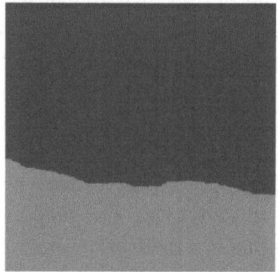 

Fig. 3. An image marked-up with "objects". This example has two groups of non-overlapping objects : one for the background, and one for the "objects" (the figures) in the foreground.

Extra "Paths" can be included to illustrate the route that objects move along in the scene being portrayed. For example, for a bouncing ball, the shape of the ball, and the path that it follows, can be presented.

A Guide can be bound to an audio file soundtrack (e.g. MP3 or WAV file). In a test, a sequence lasting approximately 150 seconds was presented via a Guide file bound to a corresponding MP3 file of acceptable sound quality. The combined file was about 500 kilobytes in size.

The system can present the most important objects and features. Alternatively the user can specify a keyword, so that only items that include the keyword in their description are presented. For each item, the system moves the "tracer" to describe the shape for the item (for example via an outline tracer, or via a symbolic tracer etc.), as well as presenting related categorical information (e.g. colours, texture etc.). The tracers can be sized to correspond to the item's size and shape; or be expanded; or expanded only when an item is smaller than a certain size.

It is found to be effective to "step" around the qualifying objects in a View, showing the most "important" objects and features (however determined), in order of importance.

For non-prepared media, the system has to look for objects to exhibit. The user can control the object selection. For example the check-boxes Fig. 4 provide a simple method of telling the system to look for particular colours. More precise parameters (e.g. specifying size, shape etc.) can be given elsewhere. Fig. 4 also shows the check-boxes for requesting that certain types of object (faces, figures, or moving objects) are looked for. (Advanced object recognition is not currently implemented for live images, but the controls could be used to select particular objects types from prepared media – for example people's faces could be requested to be presented.) Object detection and identification is not a main focus of the project as it is a major separate area of research, but simple blob-detection methods are currently implemented, and in future standard face-detection facilities etc. [7,8] may be included.

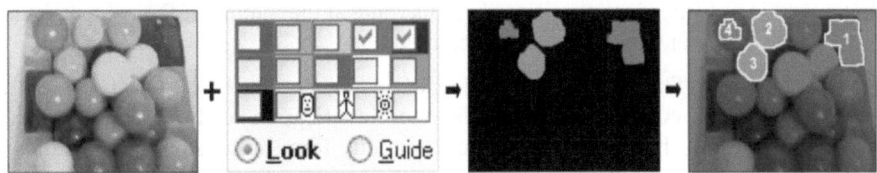

Fig. 4. Finding four blue or green objects, and presenting them in size order

Any found objects can be presented as audiotactile effects in the same way as if they had been marked-up in a prepared image – though the system has to decide which of the found objects (if any) are presented (i.e. which objects best match the user-controlled parameters), and their order of importance (e.g. by order of size).

3.2 Object Tracer Paths

The object tracer paths can follow several different types of route, and these are described below and illustrated in Fig. 5.

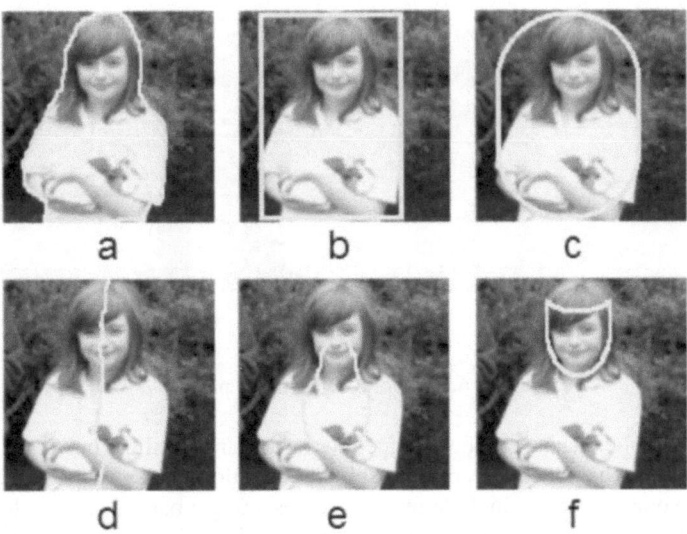

Fig. 5. "Object" tracer path types

The outline (a) of an "object" can be presented, as previously described. Alternatively the audiotactile tracer can follow a path that "frames" the extent of the object. The frame can be rectangular (b), or be rounded at the corners (c), and sloped to match the angle of the object.

The tracer can follow the "centre-line" of an object (d). This is most effective for elongated objects, where the path travels in the general direction of the longest edge, but is not as effective for objects with no clear elongation : for them, a "circuit medial" (e) can be used, where the path travels in a loop centred on the middle of the object, and is positioned at any point along its route at the middle of the content found between the centre and the edge of the object.

Symbolic Object Tracer Paths. For identified objects, the system can present a series of lines and corners that symbolise the classification of those objects, rather than presenting the shapes that the objects currently form in the scene. Fig. 6 shows example symbolic paths. Human figures (a) and people's faces (b) are examples of entities that can be effectively presented via symbolic object paths. It is best if the paths are such that they would not normally be produced by the outline of objects, for example by causing the paths to travel in the reverse direction to normal.

Currently, symbolic object tracer paths would mainly be displayed for prepared material. However image processing software can at the present state of development perform some object identification, for example by using face-detection methods [7,8]. In such cases a standard symbolic shape Fig. 6 (b) can be presented when the corresponding item is being output. An "X"-shaped symbolic object path representing "unknown" is provided Fig. 6 (c), allowing "unidentified" objects to be handled in the same way. (Alternatively the system could revert to presenting the outline or other shape when an unidentified object is processed.) Symbolic object paths are generally angled and stretched to match the angle and aspect ratio of the object being presented.

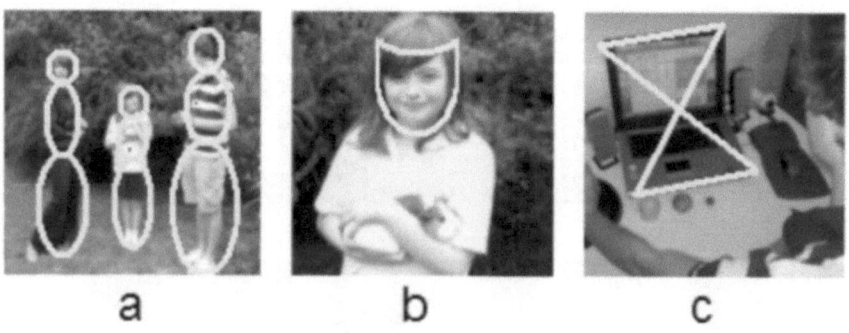

a   b   c

Fig. 6. Symbolic object tracer paths

Basic symbolic shapes can be assigned to particular classifications/types, and embellishments can be added to represent sub-classifications e.g. a shape representing a face can be embellished to include features representing a pair of glasses, left-profile, right-profile etc. by having additional effects added. By using this approach, basic symbolic shapes of common object classifications can be easily recognised by beginners, with sub-classifications recognised by more experienced users. It was found to be useful to have sub-categories of symbolic shapes that show parts of an object. For example it is useful to provide a shape for the the top half of a human figure, head and shoulders etc., as these are often what is present in a visual image.

3.3 Object-Related Layouts

When presenting objects, a "layout" related to the object can be presented at the same time, for example by using a braille display.

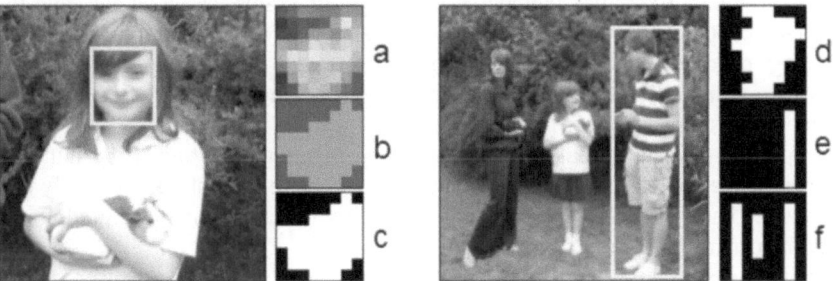

Fig. 7. Object-related layout types

Because the shape of the object is known, the image content in only the area covered by the object can be presented, spread to fill the layout area. Alternatively the content of the rectangular frame enclosing the object can be presented, with the content stretched if necessary in one direction to use the full height and width of the frame Fig. 7 (a & b).

Alternatively, the content of the frame can be presented using an approach which incorporates the perceptual concept of "figure/ground" i.e. the effect whereby objects are perceived as being figures on a background. If one object is being presented then the system can present the layouts as showing the regions covered by the object within the "frame" enclosing the object (optionally stretched to match the layout dimensions) Fig. 7 (c & d); or the location of the object ("Figure") within the whole scene ("Ground") can be presented (e) - when the system is "stepping" round the image, presenting the selected objects, the highlighted objects within the layout will appear and disappear as the corresponding objects are presented, giving the user information about their location, size and colour etc. (Alternatively all of the objects being presented within the whole scene can be displayed simultaneously Fig. 7 (f).)

If object types can be identified, then "Symbolic Layouts" can be presented (using a similar approach to that used for Symbolic Object Paths), wherein the arrangement of dots is constant for particular object types (as previously reported [9]).

When objects of particular colours are being looked for and presented, and "framed" layouts are being used (i.e. not the whole image), the frame can be set wider than the exact extent of the frame enclosing the found object, otherwise the typical effect is for the layout to show mainly the found colour. By setting the framing wider, the context in which the found colour is located is also presented.

Layouts that are output as speech or Morse (i.e. not braille) tend to be long-winded. If object-related layouts are being presented, a compact format can be used : only the location of the centre of the object can be presented, via a single "CV" syllable, the C(onsonant) and V(owel) giving the vertical and horizontal "coordinates" of the centre of the object. Additional coded "CV" syllables can give the approximate size and/or shape, colour etc. of the object if required.

3.4 Processing Simple Images

It is important that the HFVE system effectively handles simple images or other visual materials containing a limited number of colour shades, and with clearly defined coloured regions. Examples include certain maps, diagrams, cartoons etc., and these are often encountered, particularly in environments where a computer might be being used (e.g. office or educational environments). Though they can be handled via the standard routines that handle any type of image, it was found to be effective to have special processing for simple images.

Simple images can be detected by inspecting pixels and testing if the number of different shades is less than a certain value. An effective way of automatically determining the "background" colour shade is by finding which colour shade predominates along the perimeter of the image.

Such images do not require special optical filtering, as "objects" are already clearly defined in them, and these can be presented. The approach works well for simple images held in "lossless" image file formats, e.g. "GIF" and "BMP" formats. For example diagrams drawn using Microsoft's Windows "Paint" program can be effectively presented in this way, or a special facility can be provided to allow shapes etc. to be rapidly drawn and then immediately presented in audiotactile format.

4 Pilot Study

A pilot study/trial with untrained users has recently commenced. The prototype Silooet software was installed on an ordinary laptop computer, and separate speakers and two types of low-cost force-feedback devices were used, namely a Logitech Force Feedback Mouse, and an adapted Microsoft SideWinder Force Feedback 2 joystick Fig. 8.

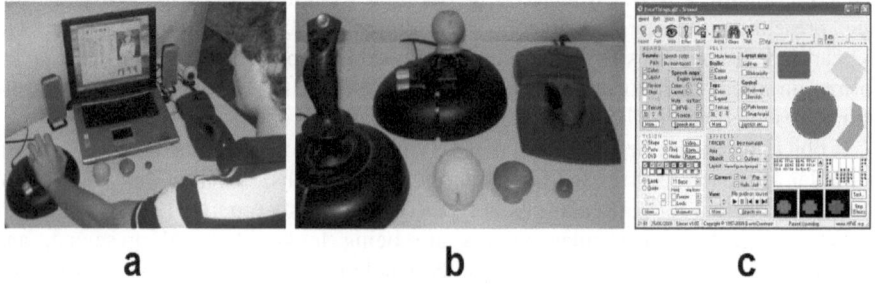

a   **b**   **c**

Fig. 8. The pilot study equipment (a); the force feedback mouse and adapted force feedback joystick, with alternative handles (b); and the main GUI for the trial Silooet software (c)

The standard joystick vertical handle configuration is designed for computer games, flight simulators etc. The handle was detached and some wires de-soldered, so that four control buttons, the slider, and potential twist action remained (the original handle could easily be re-fitted if required). Three alternative wooden handles (roughly the size of an apple, a door-knob, and a large grape) Fig. 8 were tested.

The two force-feedback devices tested in the study are low cost but not current (though relatively easily obtainable) - testing the system with current force-feedback hardware may be worthwhile. Braille output took the form of simulated braille cells shown on the main GUI, which is obviously not suitable for blind testers (a programmable braille display has not been implemented at the time of writing).

The system was presented to several sighted participants, of different ages, in informal trial/interview sessions, and their impressions were noted and performances assessed. Standard shapes were presented at various speeds and sizes, in audio, tactile, and combined audiotactile format. Sequences of filed and live images were tested.

The initial findings are:-

- After a few minutes practice, participants could typically recognise more that 90% of standard demonstration shapes. Shapes with many features, representing more complex objects, were more difficult to interpret directly, but could be recognised after several repetitions, when interspersed with the standard shapes.
- Emphasised corners are essential for recognising shapes.
- Standard shapes could be recognised at Full-, 1/2-, and 1/4- image diameters with no difficulty. Recognition became more problematic at 1/8 diameter sizes. This finding suggests that small shapes should be automatically enlarged, perhaps with audiotactile cues given to indicate that this has been done.
- Of the two haptic devices tested, their was no clear preference. The force feedback mouse had a more suitable handle configuration for the HFVE Silooet application and gave very accurate positioning (but users needed to hold it lightly by their fingertips), while the joystick gave more powerful forces. All participants preferred one of the replacement joystick handles to the standard vertical handle : the door-knob handle was preferred by a child, while older testers preferred the apple-sized handle. A tennis ball cover was added to this to provide a softer surface, and this was the most preferred joystick configuration. (The standard joystick handle was usable, but not as effective for presenting shapes.)
- Audiotactile output (i.e. both modalities together) generally worked best. Audio (speech) was most effective for categorical information, and tactile was most effective for comprehending shapes.
- None of the testers liked the Morse-code like effects (either audio or tactile "taps"), but this could be due to their lack of familiarity with Morse. The speech-based categorical effects and braille layouts are more immediately accessible.
- A "novice" mode was requested, wherein the colours (and recognised objects) are not coded, but spoken in full (this was what was originally planned [9]).
- Sudden moves of the joystick, when it was repositioning to present new objects, were found distracting by certain testers, but others felt it gave a clear indication that a new object was being presented. Some clarification via audiotactile effects is needed, perhaps with several styles of repositioning being made available.

- The most liked features were recognising standard shapes; corners; symbolic tracers; using the system to find things in live images; and the "quick draw" / simple image feature.
- The least liked features were Morse-style output, and very small shapes.

The trial/interview sessions lasted about an hour each. Participants reported feeling tired after this period, though that may have been due to the intensity of the sessions and their unfamiliarity with the system. The effects of longer-term use of the system has not yet been assessed.

At the time of writing testing has only recently commenced, and all of the testers have been sighted. It is hoped that fuller results, and the feedback of blind testers, can be reported at the HAID workshop.

5   Conclusions and Future Development

The HFVE system has now been developed to the point where the prototype Silooet software is being tested in a pilot study. The system's aim of effectively presenting the features of successive detailed images to blind people, is challenging. Some users might only use the system for accessing more straightforward material.

Future developments can build on the trial results, and attempt to create a useful application. The test done so far show that most people are able to easily recognise standard shapes. The positive response to recognising shapes, to symbolic object shapes, and to live images suggests that a future development could be to incorporate automatic face-recognition and other object-recognition facilities.

Possible applications include:- presenting shapes, lines, maps and diagrams for in-structional purposes; providing information to users wishing to know the colour and shape of an item; and for specific tasks such as seeking distinctively-coloured items. The recently-commenced pilot study should help to clarify which aspects of the system are likely to be the most useful.

References

1. Dewhurst, D.: An Audiotactile Vision-Substitution System. In: Proc. of First International Workshop on Haptic and Audio Interaction Design, vol. 2, pp. 17–20 (2006)
2. Dewhurst, D.: "Silooets" - Audiotactile Vision-Substitution Software. In: Proc. of Third International Workshop on Haptic and Audio Interaction Design, vol. 2, pp. 14–15 (2008)
3. U.S. Patent Appl. No. US 2008/0058894 A1
4. Fournier d'Albe, E.E.: On a Type-Reading Optophone. Proc. of the Royal Society of London. Series A 90(619), 373–375 (1914)
5. Vision Technology for the Totally Blind, http://www.seeingwithsound.com
6. iFeelPixel, http://www.ifeelpixel.com
7. Viola, P., Jones, M.: Robust real-time object detection. In: IEEE ICCV Workshop on Statistical and Computation Theories of Vision, Vancouver, Canada (2001)
8. Yang, M., Kriegman, D., Ahuja, N.: Detecting Faces in Images: A Survey. IEEE Transactions on Pattern Analysis and Machine Intelligence 24(1), 34–58 (2002)
9. The HiFiVE System, http://www.hfve.org

Configurable Design of Multimodal Non Visual Interfaces for 3D VE's

Fabio De Felice, Giovanni Attolico, and Arcangelo Distante

Institute of Intelligent Systems for Automation – Italian National Research Council,
Via G. Amendola 122 D/O, 70126, Bari, Italy
{defelice,attolico}@ba.issia.cnr.it

Abstract. 3D virtual environments (VE) require an advanced user interface to fully express their information contents. New I/O devices enable the use of multiple sensorial channels (vision, hearing, touch, etc.) to increase the naturalness and the efficiency of complex interactions. Haptic and acoustic interfaces extend the effective experience of virtual reality to visually impaired users. For these users, a multimodal rendering that matches the subjective characteristics and the personal abilities of individuals is mandatory to provide a complete and correct perception of the virtual scene. User feedbacks are critical since the design phase. This paper proposes an approach for the design of haptic/acoustic user interface to makes up the lack of visual feedback in blind users interaction. It increases the flexibility of the interface development by decoupling the multimodal rendering design from the VE geometric structure. An authoring tool allows experts of the knowledge domain (even without specific skills about the VE) to design the haptic/acoustic rendering of virtual objects.

Keywords: Haptic\acoustic interface design, cooperative design, visual impaired users, 3D virtual environments.

1 Introduction

3D virtual environments (VE) increase the amount of data that can be conveyed to the user but requires more complex interaction paradigms compared to the traditional WIMP approach. New I/O technologies, such as 3D pointing devices, spatial audio and haptic devices, provide tools to cope with this greater complexity through natural human/machine interactions. These devices address specific sensorial channels (tactile, hearing, seeing, ...) and propose different interfaces, see [1] for a broad review. In such multimodal interactions information is rendered in a polymorphous and redundant way: therefore each user can select the interaction modalities best suited to his own characteristics. These new approaches to VR user tasks require new interaction metaphors to be used in the design phase.

Multimodal 3D human-machine interaction enables the development of applications addressing users with sensorial disabilities, who can select the modality (haptic, acoustic, visual, gestures) that best fits their needs and personal characteristics. In particular [2][3][4] proved that haptic\acoustic VE applications represent valid tools for visual impaired users: they provide the access to and facilitate the manipulation

M.E. Altinsoy, U. Jekosch, and S. Brewster (Eds.): HAID 2009, LNCS 5763, pp. 71–80, 2009.
© Springer-Verlag Berlin Heidelberg 2009

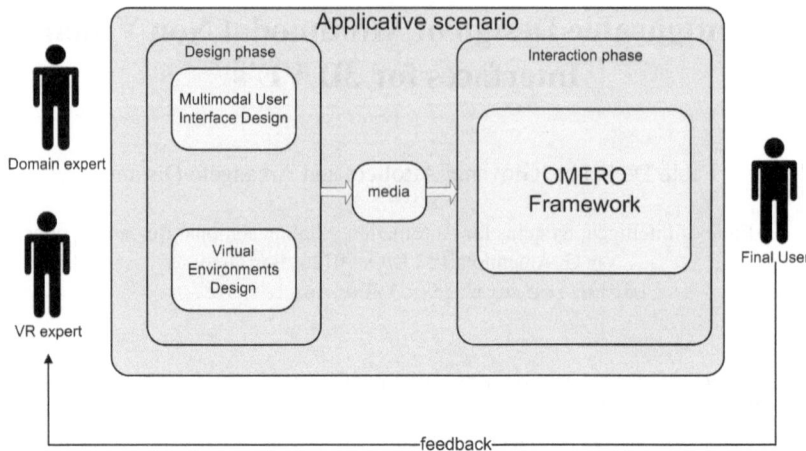

Fig. 1. The applicative scenario

and comprehension of many kinds of information usually conveyed by physical arte-facts that are sometimes less effective and more expensive.

Our previous research [4] [5] lead us to envision an applicative scenario, shown in figure 1, in which domain experts and VR experts design and create VEs and their multimodal interfaces using properly developed tools. A multimodal system, called OMERO [4], allows visually impaired users to explore such 3D scenes. The design of the user interface becomes a challenging task because haptic and acoustic have to compensate for the lack of visual feedback. This means that haptic and acoustic ren-derings of 3D objects have to emulate some important features of vision. Moreover, blind users show a wide variability in terms of subjective characteristics and sensibil-ity: expressing effectively the intended informative content requires a careful consid-eration of users feedback to refine design choices.

The presented approach decouples the design of the VE from the design of its hap-tic\acoustic interface. In this way a one to many relationship can be established be-tween a VE and its possible user interfaces. To offer a wide variety of metaphors as long as a large flexibility in matching the rendering to specific user needs, tools to design and modify quite easily and quickly the multimodal user interface are pre-sented. The VEs design and development is not take into account because many tools exists to create VEs, from authoring tools (Google SketchUp) to complex CAD sys-tems (Maya, AutoCad,...).

Such a decoupling generates two related outputs: a VRML structure describing the geometry f the scene and a XML based structure containing the user interface. Such XML based schema called OMDL (Omero Multimodal Markup Language) allows an easy association of a particular interface\display combination with the virtual world. An editor tool (Editor herein) allows a visual authoring of the OMDL file on a specific virtual scene. The designed VE can be successively explored by blind users using the OMERO multimodal framework. The users feedbacks provided by the final users during the experience can be used by domain experts and VR engineers to refine the design choices by using the Editor, in an iterative and adaptive application lifecycle.

The paper is organized as follows: in section 2 the multimodal metaphors used for define the conceptual framework behind the multimodal user interface are presented and compared with analogous approaches used in applications of VEs for sighted users. In section 3 the OMDL schema is described and the Editor is presented and its features and functions are described. In section 4 an application with blind users is reported and, finally in section 5 conclusions are drawn.

2   Non Visual 3D User Interface

We have focused our attention on a set of haptic/acoustic interaction metaphors for the design multimodal user interface for visually impaired users. There are no universally accepted standards for this task, although a lot of research has been carried out and few guidelines have been proposed [6][7].

Our interface implements previously proposed guidelines and adds two new interaction techniques: Active Objects and Scenarios. Our conceptual framework supports different user tasks: navigation, objects selection\manipulation [8] and what could be called scene querying. As stated in [9], not all the existing metaphors of interfaces with VEs can be suitably enhanced by haptic. Despite this, haptic and acoustic have to make up for the lack of visual display in a multimodal interaction addressing blind users.

- **Navigation:** metaphors for visual navigation mainly rely on camera manipulation methods [1][9], which can be roughly divided in egocentric and exocentric techniques. Egocentric camera manipulation require a continuous change of the point of view depending on the user movements: this can confuse blind users that miss stable references points [6]. Furthermore egocentric navigation approaches for visual impaired are poorly documented (an example can be found in [10]). For these reasons, exocentric metaphors are mainly used.

In our approach the movement of the haptic stylus does not affect the reference system of the VE, as with a (Haptic) World in Miniature metaphor [9], that keeps constant absolute position and orientation with respect to the desk. The avatar moves in a virtual workspace delimited by a **Containment Box** directly coupled with the physical dimension of the haptic device workspace. The user explores the VE in a "God-eye" way. **Acoustic messages** inform the user whenever (s)he enters new regions of the scene (a useful cue to support wayfinding [1]). The **Containment Box** prevents blind users from moving too far from the virtual scene. This feature proved to be a valuable help to prevent the user from wasting time in useless regions of the workspace, getting back in touch with the objects of interest.

To face the problem of the limited physical workspace **Dragging** and **Scaling** functionalities are provided, based on the work in [11], to dynamically change which part of a large model is shown in the workspace and its scale. Two types of dragging are given: in the **Stylus Dragging** technique the user has the impression of moving the whole VE model (according to the movements of the stylus) in respect of the position of the containment box. In the **Box Dragging** technique the user has the impression of moving the containment box (by pushing on its walls) in respect of the VE. The **Scaling** technique dynamically changes the sizes of scene details according to the dimension of the user avatar. This can reduce the degree of manual dexterity required for their correct perception. If the user requires the scaling while touching an object, the VE is scaled according to the position of this contact point, otherwise the scaling is made according to the position of the centre of the scene. This meaningful reference prevents the user from being confused by an uncontrolled movement of the environment. Preliminary experiments with blind users showed that choices of the more comfortable method seemed to be influenced by subjective user characteristics and that scaling is well accepted and used in combination with dragging to navigate in particular regions of the scene. However further studies are needed to better ground these conclusions.

Metaphors such as "Teleportation" [12] and "Small Scene Manipulation" [13] can both be used in visual interaction as basic techniques to move the user avatar between two targets (target-based travel) and for route planning [1]. The first proved to confuse sighted persons [14], and force feedback does not provide any improvement for this type of users. Contrariwise suitable attractive forces [6] can be applied to the blind user hand to guide the exploration: in our approach a sequence of targets defined by the designer [14] form a **Guided path** which guides users through salient regions of the scene. Vocal messages inform the user that the target has been reached. This haptic\acoustic rendering proved to increase the completeness and effectiveness of the mental schema constructed by blind users during the virtual environment exploration. This kind of guided tours (or haptic glance) supports wayfinding allowing a first coarse perception of all the relevant features of the model.

- **Object selection and manipulation:** this task require objects in the scene to be suitably organized and associated with haptic\acoustic displays, to facilitate their perception by blind users. We defined **Active Objects,** [4] as parts of the scene conveying further information apart from their shape and geometry. These objects are associated with specifically defined actions when touched by the user. Haptic effects such as attraction, vibration, or viscosity can be used to identify the associated objects. Active Objects can be classified as haptic, acoustic or haptic/acoustic according to their associated renderings effects (haptic, acoustic or some combination of both modalities). Active Objects can be associated with a vocal message providing additional information (i.e. historical or artistic descriptions, dimensions, material, etc.). Other type of objects will be referred to as background objects. Some active objects (referred to as dynamic) that have a function in the environment, such as doors, windows or switches, hare associated with translation\rotation predefined behaviors that the user can trigger. In this way components of the scene that are relevant for its comprehension can be properly emphasized.

To adapt the interaction between the user and the active objects to individual needs, effects and dynamic behaviors can be produced under two different conditions: **On Touch**, that is automatically whenever the user gets in touch with the Active Object. **On Demand**, only if the user asks for it, in order to avoid an unintentional and sudden scene modification.

Object selection task in visual interfaces exploit metaphors such as Ray-Casting and Cone-Casting [9]. A type of haptic metaphor similar to the Cone-Casting could be the Navigation Sphere in [7]. A metaphor similar to Ray-casting is used in our work: attractive force fields surrounding Active Objects. When the avatar is inside an attractive field, the stylus is attracted, automatically or on demand, towards the object whose nature is specified by a vocal message. This technique has been successfully used to make users aware of Active Objects placed in the central part of the environment (less probably visited by blind people that tend to move along the borders of spaces): from this point of view it can be seen as both a selection and a navigation technique.

- **Scene Querying:** A complex virtual world, rich of several kinds of details, generates a long sequence of local perceptions. Integrating all these data into a coherent and meaningful mental schema is often a real challenge. A VE can be seen as a database containing different type of information represented in the form of 3D objects such as different kinds of data available in a geographical map or different anatomical systems of an organism. From this point of view a user can explore a VE querying the scene for different type of information to be displayed. We organize the scene in **Scenarios,** views that represent semantically consistent and coherent parts of the complete information content of the scene. A Scenario can be turned on and off (being touchable and visible or not) and is a set of semantically related Active Objects. This is accomplished by inserting Switch nodes in the VE scene graph opportunely. The user, at each specific time, can query the scene asking for a particular Scenario and focus his\her attention only on the information associated with it, temporarily discarding all the other data. The perception of the scene can therefore be tailored to focus attention on the data of interest in a progressive access to information.

3 Authoring the Scene

In this context the design work focus on how to configure an instance of the conceptual framework previously described. To do this all the metaphors are been integrated together in a unique model. The scene is viewed as a series of Scenarios, each offering a specific semantic view of the whole information content. Each scenario may have a "global" effects associated with it, which is active all the time except when a "local" effect, triggered by an Active Object, is turned on. Furthermore, each Scenario may be associated with a guided path. An object can belong to several Scenarios and Active Objects can belong to different guided paths. Whenever needed the user can drag and scale the scene.

Our approach decouples "what" must be rendered (the VE geometry and structure) from "how" it is proposed to the final blind user (the multimodal interface), because there is not a unique optimal choice of effects to convey a specific information. Even the subdivision of the scene into Scenarios may change over time depending on the specific communication goals and on the target users. In this way a one to many relationship can be established between the VE and its possible interactions with the user. Therefore, the design process must be open-ended: every rendering of a specific scene should be improved and personalized as soon as new user feedbacks become available. Moreover, information contents are generally decided by the domain expert, which is not necessarily well grounded in 3D multimodal design. This suggests that the structure of the user interface should be authored directly by those who are responsible for the scene contents. To account for these requirements an intermediate human-readable media is needed to describe an instance of the multimodal user interface: an XML schema called OMDL (Omero Multimodal Description Language) has been created to allow designer to easily configure the interface, adding sharing and reuse potentialities to the approach.

```
<world>
  <haptic>
    <device>PHANToM Omni</device>
    <API>OpenHaptics</API>
  </haptic>
  <scene>
    <scenarios>
      <scenario id="0" text="scenario 1: Forme">
        <globalEffect>
          <globalHapticEffect>
            <code>gravity</code>
            <triggerMode>On Touch</triggerMode>
          </globalHapticEffect>
        </globalEffect>
        <guidedPath>
          <target id="0" type="object">Obj1</target>
          <target id="1" type="object">Obj2</target>
          <target id="2" type="object">Obj3</target>
          <target id="3" type="3Dpoint">12.23415,-10.345678,56.098</target>
        </guidedPath>
        <switchConfiguration>
          <switchValue id="0">NONE</switchValue>
          <switchValue id="1">NONE</switchValue>
          <switchValue id="2">ALL</switchValue>
          <switchValue id="3">ALL</switchValue>
          <switchValue id="4">3</switchValue>
          <switchValue id="5">NONE</switchValue>
          <switchValue id="6">ALL</switchValue>
        </switchConfiguration>
      </scenario>
```

a)

```
<iObject id="0" name="kitchenDoor" type="haptic/acoustic">
  <material>
    <stiffness>0.8</stiffness>
    <friction>0.3</friction>
  </material>
  <behavior>
    <localEffect>
      <hapticEffect>
        <code>vibration</code>
        <triggerMode>onTouch</triggerMode>
      </hapticEffect>
      <acousticEffect>
        <code>door</code>
        <triggerMode>onDemand</triggerMode>
        <acoustic id="0">openDoor.wav</acoustic>
        <acoustic id="1">closeDoor.wav</acoustic>
      </acousticEffect>
    </localEffect>
    <tts>
      <string>The kitchen door</string>
      <triggerMode>on demand</triggerMode>
    </tts>
    <dynamic>
      <translation id="0">0.0,10.0,0.0</translation>
      <translation id="1">0.0,-10.0,0.0</translation>
    </dynamic>
    <eventsSequence>
      <event type="acoustic" id="0"/>
      <event type="translation" id="0"/>
      <event type="translation" id="1"/>
      <event type="acoustic" id="1"/>
    </eventsSequence>
  </behavior>
</iObject>
```

b)

Fig. 2. Examples of the OMDL schema. In a) the Haptic and Scenarios description section b) excerpt from the description of an Active Object.

The schema is divided in three sections, the first section is about the chosen haptic framework (figure 2.a): since the multimodal system OMERO is independent from the haptic device used, data about which device to adopt must be supplied to configure the system. The second section describes data for all the defined Scenarios (figure 2.a). For every Scenario data about global effects, guided paths and configuration of Switch nodes are given. The Switch grouping nodes select which of their children will be visited during the rendering by indicating one of the following values: none, all or one of their children. During scene exploration, activating a Scenario means to set these values for every Switch in the scene graph according to the decisions taken during the design phase. The last section concerns Active Objects (figure 2.b). For every Active Objects there are data about the material (stiffness and friction), local effects (haptic, acoustic or both) and relative trigger mode (on touch or on demand), eventual TextTo-Speech string and the associated trigger mode, a series of dynamic events that can be 3D translation and/or 3D rotation, a sequence of events (dynamic or acoustic) that occur in the time line during the object motion and finally the Scenario the object belong to.

To allow an easy and straightforward configuration of the OMDL file an Editor has been developed whose layout is depicted in figure 3. It uses contextual menus and 2D UI to allows domain experts to load a particular VE and to develop the relative user interface. The output of the application is the OMDL file that is loaded by the OMERO framework together with the VRML file describing the VE.

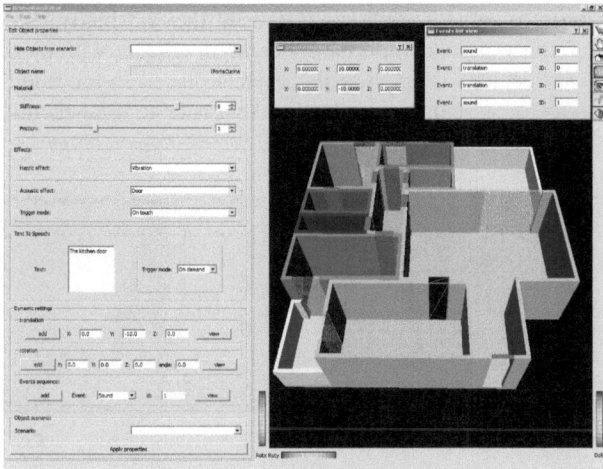

Fig. 3. The Visual Editor look and feel

3.1 The Editor

Once loaded the VRML scene into the Editor two main tasks can be performed: defining Active Objects and creating Scenarios and their guided paths. The former task requires an object in the scene to be selected using the mouse. A contextual menu (the one on the left in figure 3) allows the designer to edit the object interface. The "Scenarios" and "Guided path" menus (not shown here) allows the designer to select an already inserted Scenario to check or modify its state and to select the belonging Active Object to form a guided path.

As an example of the above setting, consider the black foreground door in figure 1: its OMDL configuration as a dynamic haptic/acoustic active object is shown in figure 2.b. The object has been associated with a stiffness value of 8 and a friction of 3: these values are normalized in the range of [0,1] when the XML is produced. A haptic effect of vibration has been chosen to distinguish the door from the walls. Consequently the haptic trigger mode has been set to "on touch".

The acoustic effects imply two .wav files (the sound of an opening and of a closing door). Because the acoustic effect is related to the opening\closing dynamic its trigger mode has been set automatically to "on demand". Two translations have been defined. The dynamic is described as the sequential trigger of: an opening door sound, a then a vertical opening translation, a vertical closing translation (opposite to the previous one) and a closing door sound.

The Editor enhances the internal structure of the VRML scene graph too. Indeed while creating Scenarios, the Editor automatically inserts proper Switch nodes in the scene graph. The insertion and configuration of Switch nodes groups Active Objects belonging to the same Scenario and close to each other under the same switch node, accounting for their position inside the whole scene. Figure 4 describes an example of the use of Scenarios. In the depicted VE two scenarios have been designed, one containing Active Object placed in the middle of the environment and another one containing Active Objects placed along the boundaries. It can be useful for a blind user to perceive these types of objects separately: indeed they normally use objects on the boundaries as reference points while hardly came in contact with objects in the middle. When the user decides to explore the objects placed in the middle of the environment, attractive forces can be used to make easier to find them inside the free space.

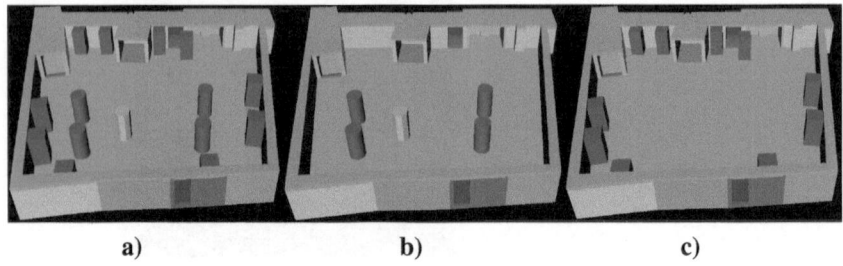

a)   b)   c)

Fig. 4. Scenarios functioning, a) a VE with all the objects inside, two scenarios have been defined: b) only active objects in the middle are rendered, c) only active objects on the boundaries are rendered

4 Preliminary Experiments

A didactical application has been designed and realized following the previously described approach and then tested with blind users with the aim of enabling the multimodal experience of basic concepts of plane geometry. Four Scenarios have been designed: Shapes, Geometric features, Perimeters and Areas. Active objects (shape, perimeters and areas) have been associated with synthetic speech that, on demand, inform the user about their dimension and related mathematical formulas. Furthermore, active objects have been rendered with a rough material in contrast with the smooth Containment Box. Active objects corresponding to geometrical characteristics have been associated with vibration effects, triggered on touch, and synthetic speech, triggered on demand.

This application has been tested on a visually impaired user that had never used haptic interfaces (a PHANToM Omni in this case) before. He expressed a positive feedback on the possibility of focusing on particular informative contents and on the autonomous switch among scenarios. Haptic rendering and effects were useful to perceive the circular sector while they did not help the comprehension of other geometric characteristics: for example it was difficult to follow the radius when its presence was highlighted only by the vibration effect. Synthetic speeches were of great help in integrating information coming from the haptic feedback with high-level information on the related concepts.

On the base of these feedbacks the scene was re-authored with the Editor: the interface of the indicated Active Objects was changed using the Active Object Menu. The vibration effect on the circular sector and on the radius was left unaltered. The diagonal of the square and the cord of the circle chord were associated with attractive force effect while some features of the pentagon and the height of the triangle were associated with a material with high friction.

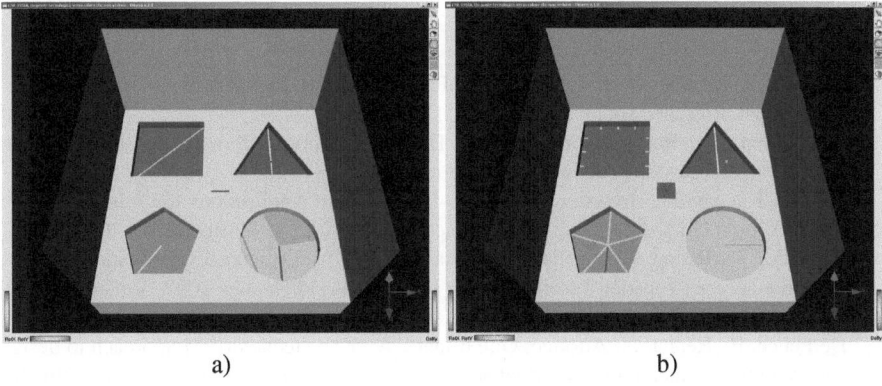

a)   b)

Fig. 5. Two of the designed application scenarios: a) Geometric characteristics scenario b) Areas scenario

The V.E. with its new interface was tested on other two blind users. Their feedbacks were similar to those of the first user. Moreover they perceived well the active objects associated with attractive forces while provided negative feedback about the use of materials with high friction.

5   Conclusion

This paper describes an approach to design multimodal 3D user interface for the interaction of blind users with VEs. This kind of design must be kept flexible: no standard hapto-acoustic interaction metaphors are universally accepted and the effectiveness of haptic/acoustic rendering for visually impaired people strongly depends on subjective factors and personal abilities. The proposed approach integrates multimodal interaction metaphors for non visual exploration of VEs: they try to make up for the lack of visual display. The approach intends to exploit user feedbacks to refine the design choices about user interface and haptic/acoustic effects used to multimodally render the scene. To reach this goal, the descriptions of the user interface and of the VE geometric structure are decoupled. The former is described by an XML schema called OMDL. In this way a single VE can be associated with a large number of different renderings tailored to specific individual user needs. Moreover the use of XML schema add sharing a reuse potentialities to the approach.

The use of a visual Editor tool to configure the OMDL file by directing authoring the VE simplifies the design phase: the most important part of the design can be made by experts of the domain and by final users, even if they don't have technical skills in the VR field. The former can contribute by selecting the relevant semantics and by organizing the structure of the scene in the way best suited to communicate those information contents. The latter can select the multimodal rendering that matches their own characteristics and sensorial limitations. Some preliminary tests conducted on three blind users suggests how this cooperative process strongly simplify design and refinement phases making the final experience effective for the intended goals.

References

1. Bowman, D.A., Kruijff, E., La Viola, J.J., Poupyrev, I.: An Introduction to 3-D User Interface Design. Presence 10(1), 96–108 (2001)
2. Magnusson, C., Rassmus-Gron, K.: A Dynamic Haptic-Audio Traffic Environment. In: Proc. of Eurohaptics 2004, Munich, Germany, June 5-7 (2004)
3. Yu, W., Kangas, K., Brewster, S.A.: Web-based Haptic Applications for Blind People to Create Virtual Graphs. In: Proc. of 11th Haptic Symposium, Los Angeles, CA (2003)
4. De Felice, F., Renna, F., Attolico, G., Distante, A.: A haptic/acoustic application to allow blind the access to spatial information. In: Proc. of World Haptics 2007, Tzukuba, Japan, March 22-24 (2007)
5. De Felice, F., Renna, F., Attolico, G., Distante, A.: Knowledge driven approach to design and multimodal interaction of virtual scenes for visual impaired users. In: Proc. of Intelligenza Artificiale nei Beni Culturali, IABC 2008, Cagliari, Italy, September 11 (2008)
6. Sjostrom, C.: Using Haptics in Computer Interfaces for Blind People. In: Proc. of CHI 2001, Seattle, USA, 31 March-5 April (2001)
7. Pokluda, L., Sochor, J.: Spatial Orientation in Buildings Using Models with Haptic Feedback. In: Proc. of World Haptics Conference, First Joint Eurohaptics Conference and Symposium on Haptic Interfaces for Virtual Environment and Teleoperator Systems (WHC 2005), pp. 523–524 (2005)
8. Gabbard, J., Hix, D.: A Taxonomy of Usability Characteristics in Virtual Environments. Virginia Polytecnic Institute and State University
9. De Boeck, J., Raymaekers, C., Coninx, K.: Are Existing Metaphors in Virtual Environments Suitable for Haptic Interaction. In: Proc. of the 7th International Conference on Virtual Reality (VRIC 2005), pp. 261–268 (2005)
10. Colinot, J.P., Paris, D., Fournier, R., Ascher, F.: HOMERE: a Multimodal System for Visually Impaired People to Explore Virtual Environments. In: Proc. of IEEE Virtual Reality Conference, March 22-26 (2003)
11. Magnusson, C., Rassmun-Grohm, K.: Non-visual zoom and scrolling operations in a virtual haptic environment. In: Proc. of the 3rd International Conference Eurohaptics 2003, Dublin, Ireland (July 2003)
12. Bowman, D., Koller, D.: A methodology for the evaluation of trave techniques for immersive virtual environments. Virtual Reality Journal 3, 120–121 (1998)
13. De Boeck, J., Cuppens, E., De Weyer, T., Raymaekers, C., Coninx, K.: Multisensory interaction metaphors with haptics and proprioception in virtual environments. In: Proc. of NordiCHI 2004, Tampere, FI (October 2004)
14. De Felice, F., Renna, F., Attolico, G., Distante, A.: A Portable System to Build 3D Models of Culturale Heritage and to Allow Their Explorations by Blind People. In: Proc. of HAVE 2005 – IEEE International Workshop on Haptic Audio Visual Environments and their Applications, Ottawa, Ontario, Canada, October 1-2 (2005)

Tactile Paper Prototyping with Blind Subjects

Mei Miao[1], Wiebke Köhlmann[2], Maria Schiewe[3], and Gerhard Weber[1]

[1] Technische Universität Dresden, Institut für Angewandte Informatik
Nöthnitzer Straße 46, 01187 Dresden, Germany
{mei.miao,gerhard.weber}@tu-dresden.de
[2] Universität Potsdam, Institut für Informatik
August-Bebel-Straße 89, 14482 Potsdam, Germany
koehlmann@cs.uni-potsdam.de
[3] F. H. Papenmeier GmbH & Co. KG
Talweg 2, 58239 Schwerte, Germany
maria.schiewe@papenmeier.de

Abstract. With tactile paper prototyping user interfaces can be evaluated with blind users in an early design stage. First, we describe two existing paper prototyping methods, visual and haptic paper prototyping, and indicate their limitations for blind users. Subsequently, we present our experiences while preparing, conducting and analysing tests performed using tactile paper prototyping. Based on our experiences, we provide recommendations for this new usability evaluation method.

Keywords: tactile paper prototyping, low-fidelity prototyping, usability evaluation method, visually impaired, tactile interaction, design methodologies, usability, user-centred design.

1 Introduction

Paper prototyping is a widely used method in user-centred design (see ISO 13407) to develop software that meets users' expectations and needs. Testing concepts with prototypes before implementation allows for inexpensive changes as paper mock-ups, e. g. hand-sketched drawings, can be adapted quickly according to users' comments.

Henry [1] describes common procedures for testing accessible software. Besides the conformance to accessibility standards, evaluation expertise and the experience of people with disabilities is needed to evaluate applications. Specified methods such as heuristic evaluation, walkthroughs and screening techniques can be conducted with design team members or users. Though, Henry does not describe any concrete methods for testing with subjects with disabilities in the early development stage.

Visually impaired access digital information using assistive technology such as screen readers and Braille displays. Within the project HyperBraille[1], a tactile two-dimensional display with the size of 120×60 pins, the BrailleDis 9000, is

---

[1] HyperBraille project website: http://www.hyperbraille.com/

M.E. Altinsoy, U. Jekosch, and S. Brewster (Eds.): HAID 2009, LNCS 5763, pp. 81–90, 2009.
© Springer-Verlag Berlin Heidelberg 2009

being developed which can be used to display multiple lines of text and graphical information [2]. Furthermore, interaction is possible through its touch-sensitive surface. In addition to the hardware, a software system for presenting content of conventional applications (e. g. Microsoft Office, Internet Explorer) is being developed which considers the special needs of blind users. The adaptation of detailed and coloured GUIs to a lower binary tactile resolution with adjusted interaction techniques required an elaborate conceptual design accompanied by ongoing formative evaluation[2].

The first usability test of our concept's tactile user interface was conducted in an early development stage. For this evaluation we applied tactile paper prototyping (see Section 3) in combination with audio confrontation [4]. The focus of this paper is not to give a summary of our evaluation, but to present our observations and recommendations for conducting tactile paper prototyping.

The paper is structured as follows. First, an overview of types of paper prototyping is given. After a brief description of the evaluation, our observations while preparing, conducting and evaluating our tests are discussed. In the following, recommendations for conducting tests using tactile paper prototyping with blind subjects are given. The paper closes with a conclusion and an outlook.

2   Paper Prototyping

In user-centred design, paper prototyping is a widely used low-fidelity usability inspection method to evaluate drafts in an early stage of product design. Prototypes similar to the final product are called *high-fidelity* while those less similar are called *low-fidelity* [5]. In this paper we focus on prototypes for user interfaces.

Subjects evaluate products or applications via mock-ups that provide low functionality and can consist of different materials. Alternatively, prototypes, usually computer applications, provide more functionality, but are normally created later in the development process when basic concepts have already been approved [6]. Paper prototyping does not only serve to evaluate existing concepts and to identify weaknesses, it also offers the possibility to the subjects to make suggestions for improvement. This technique is very inexpensive and effective as it allows for testing a product before implementation.

Mock-ups are prepared in advance by *mock-up designers* who are not necessarily identical to the product designers. Conducting paper prototyping normally requires four responsibilities: greeter, facilitator, computer and observer [7]. A greeter is responsible for welcoming subjects, the facilitator conducts the session. Usually developers play the role of computers, manipulating the interface pieces according to the subject's actions. Observers take notes and are responsible for recordings.

Depending on modes of perception, we can classify current paper prototyping methods in visual (2.1) and haptic (2.2) paper prototyping. According to this naming convention we call our new approach tactile (2.3) paper prototyping.

---

[2] For more information on formative evaluation see [3].

2.1 Visual Paper Prototyping

Among the three methods visual paper prototyping is the most widely used. Its mock-ups consist of drawn interfaces on one sheet of paper or of several movable and interchangeable individual interface elements which simulate interaction on a static background interface.

According to Snyder [8] most paper mock-ups do not need straight lines or typed text and consistent sizing of components. A complete and neat looking design rather encourages unwanted pedantic feedback, e. g. concerning alignment and sizing. Additionally, Snyder points out that paper prototyping encourages creativity as the handwritten mock-up looks unfinished.

The nature of visual paper prototyping assumes that subjects are sighted and can evaluate the designs with the help of visual information. Thus, this method excludes visually impaired and blind users.

2.2 Haptic Paper Prototyping

Haptic paper prototyping is a special form of haptic low-fidelity prototyping [9]. It serves to simulate and evaluate haptic interaction with a haptic application in an early development stage. A common material for mock-ups is cardboard.

In contrast to visual paper prototyping, haptic paper prototyping can find limited use with visually impaired and blind subjects under the condition that pure haptic interaction is concerned and visual perception is unnecessary. For this reason this method is seldomly applicable with blind subjects when evaluating GUIs. One of the few examples is a media set (see Fig. 1) for teaching graphical user interfaces to blind students, developed by the project EBSGO [10].

Fig. 1. GUI-Taktil of the EBSGO project showing a search dialog box

Moreover, Tanhua-Piironinen and Raisamo [11] used two types of haptic mock-ups consisting of cardboard models and plastic artefacts for tests with visually impaired children. They pointed out that a possible drawback of this method was the abstract model which does not allow for a full conception of the application as a whole.

2.3   Tactile Paper Prototyping

To be able to perform tests with blind subjects, we adapted visual paper prototyping according to our requirements. In principle, the concept (see Section 2.1)
is also applicable to test user interfaces with blind subjects, but the special needs
of this user group have to be considered.

Accordingly, our evaluation method allows for evaluating user interfaces re-
specting the subjects' needs. Indeed, speech output of screen readers can be
presented neither by paper nor by tactile mock-ups. Thus, to create a realistic
work environment, speech output needs to be presented by the conductors in
the role of computers.

The integration of haptic prototyping techniques in our method is conceivable,
as interface elements with a structured surface can be used to indicate certain
details, e. g. focused elements, as a compensation for highlighting on visual mock-
ups. In our special case, evaluating an interface for a device whose pins can only
be set or not (on/off), additional haptic elements were not needed.

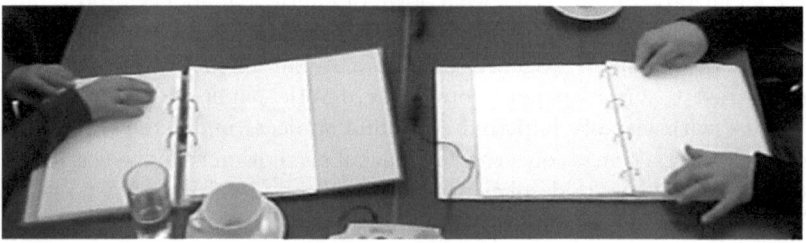

Fig. 2. Subjects exploring our tactile mock-ups

Test material for visual paper prototyping usually consists of individual interface elements which can be produced, arranged and changed quickly according
to the simulated action or the suggestions of the subject. As more preparation
is needed for tactile mock-ups when compared to the pen-and-paper approach,
changes are unlikely while conducting the evaluation. Anyhow, it is possible
to generate new mock-up elements with users during the evaluation e. g. using
paper and heat-pen or Braille paper.

3   Evaluation

The evaluation is only briefly overviewed as the focus of this paper lies in the
evaluation technique of tactile paper prototyping. A comprehensive description
of the evaluation can be found in [4].

In our evaluation, we used paper prototyping in a vertical[3] and low-fidelity
manner with a total of 11 blind subjects in groups of two to five. Low-fidelity

---

[3] Vertical prototyping tests the exact functionality of few elements of a GUI. In contrast, horizontal prototyping tests a broad spectrum of GUI elements with a low
level of functionality.

prototyping was appropriate as it was the first test of the designed interface. We used vertical prototyping, as we only focused on some features such as layout and application concept, while neglecting navigation and interaction.

Our test comprised several scenarios with one or two pages each which represented the adapted GUIs for our two-dimensional tactile device. Embossed printings, matching the resolution of the target output device, served as test material. Within our project we relied on two self-developed programmes called HBGraphicsExchange and HBBrailleExchange [12] which allowed our blind and sighted designers to create mock-ups in the appropriate resolution and size, printable with any embosser. To allow for comfortable turning and to sustain the order, the sheets were assembled in binders (see Fig. 2). The hands of the subjects on the mock-ups were videotaped and the discussions were recorded.

4   Observations

In the following, we only report on observations concerning blind subjects. Of course, most general aspects for preparing, conducting and analysing a test with sighted subjects [3] apply as well. Additionally, general issues [1][13] regarding blind subjects have to be considered. Subjects might need escort and transportation to and from the facility and are likely to bring along a guide dog. Thus, setting up the test takes more time and during its conduction additional breaks might be needed. Furthermore, extra room for service animals or assistants is needed and obstacles should be cleared out of the way.

4.1   Preparing

For tactile paper prototyping thorough preparation is important. To be able to present several solutions, it is advisable to produce multiple mock-ups for each scenario in advance, as new ideas of subjects are difficult to cover during a test.

Embossers. Creating tactile paper mock-ups requires special hardware. In the following we only discuss Braille printers, called embossers, because their prints have similar physical properties as Braille displays and most planar tactile devices. In general, the material used should resemble the final product as much as possible to ensure realistic results.

Before designing tactile mock-ups, essential facts of the embosser are needed. These comprise the embosser's resolution, equidistant or non-equidistant output, and the format and type of paper needed. Printable file formats of the chosen embosser affect the choice of the software for designing the mock-ups.

Software. Alternatively to the programmes for creating mock-ups used in our evaluation, Word documents can be interpreted by most printer drivers. While this works well for text, graphics are interpolated, resulting in tilted lines of varying thickness. Thus, it is best to create graphics matching the embosser's resolution. Graphics software can also be used, if its file formats are directly supported or if translation software is available.

Material. In contrast to sighted users who perceive representations as a whole and focus on details later on, blind users first explore details to construct a complete mental model. Thus, one representation of our test comprising multiple widgets on one sheet of paper caused difficulties in locating the described element. Therefore, it is advisable to display only one representation per sheet.

To avoid orientation difficulties due to sketchy representations it proved essential to map geometrical shapes as precise as possible, and to maintain the proportions and scale of the original output device, especially as Braille requires a fixed resolution. Thus, the special needs of the target group impede conformance to Snyders demand for handwritten visual mock-ups (see Section 2.1).

In our test, subjects were irritated by missing elements as some regions of our representation were not as detailed as the planned implementation. In such cases missing elements need to be indicated on the mock-ups or mentioned by the facilitator to avoid confusion. As it is more difficult to refer to and to discuss certain elements or positions on the mock-ups than with sighted subjects, a coordinate system can be helpful to improve the subjects' orientation, e.g. dividing the representation like a chess board and referring to the squares.

Proofreading. Before producing copies for the subjects, mock-ups should be proofread by Braille literates. This is necessary as spacing and spelling mistakes frequently occur when transcribing from print to Braille. In a non-visual context, spelling mistakes irritate even more than in visual mock-ups as the two-dimensional representation is unfamiliar to the users and an overall overview is missing. Additionally, the facilitator, not necessarily Braille literate, cannot control and correct mistakes as easily as in print.

Unless sighted designers are experts in Braille and working techniques of the blind, it is, of course, best to include blind mock-up designers in the preparation process. A couple of our mock-ups were dismissed by the subjects because the design strongly resembled visual concepts. These mock-ups had not been reviewed by our blind designers and had inevitably failed during the test.

Assembling. Supporting a smooth work flow, the order of the mock-ups used has to be consistent with the tasks. It proved useful to label the sheets with numbers or letters in Braille and print for the subjects and the facilitator.

When movable individual interface elements are used, they have to cling to the main representation (with help of magnets, sticker, felt etc.) to avoid shifting during exploration. As we conducted our tests with groups, individual interface elements would have been impracticable. Thus, changing screens were simulated by different prints, available by turning the page.

4.2 Conducting

After preparing the test material, five main aspects have to be taken into consideration. These comprise setup, recording and team structure. Furthermore, timing and explaining are crucial for tactile paper prototyping.

Setup. Our mock-ups were in landscape format; therefore the opened binders faced the subjects with their narrow side. In one location, the width of the table was not sufficient, thus the two facing binders touched each other and impeded the turning of pages. It is favourable to choose a table which is large enough to arrange all test materials comfortably and to allow the subjects to move their hands and arms freely to avoid collisions while exploring a paper mock-up.

Recording. As a repetition of tests with small user groups is difficult, it is advisable to prepare and record the sessions carefully. In advance, the positions of microphones and cameras need to be considered and tested depending on the seating arrangements and the area of interest to be recorded. We positioned a camera on the table to record the subjects' hands and the mock-ups.

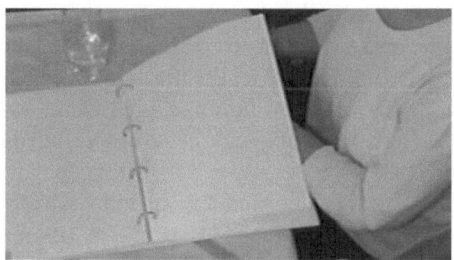

Fig. 3. Subject with hands under mock-ups

Unfortunately, with some subjects the recording was useless, as they explored the mock-ups with their hands underneath the previous sheet of paper, i. e. they did not turn the page (see Fig. 3). In such cases the facilitator must ask the subjects to turn the pages completely.

While the video shows, where the subjects' hands are positioned, it does not show satisfactorily if there was contact with the mock-up at all and which parts of the hand touched the mock-up, and how much pressure was applied.

Team. When using movable individual interface elements, the number of people needed for conducting tactile paper prototyping increases with the number of subjects. For a test with several subjects, the facilitator cannot demonstrate tactile representations to all group members simultaneously. Therefore, subjects must be provided with their own copy of a mock-up as it can only be explored by one person at a time. The subjects must be able to handle these copies on their own or, depending on complexity, one assistant (in the role of the computer) for one or two subjects is necessary.

Timing. Before dealing with a new mock-up, the facilitator has to make sure through announcing the mock-up's label that each subject has the correct material in front of him. Furthermore, the facilitator has to consider different reading speeds and exploration styles (one- or two-handed) of the subjects. Without allowing for sufficient exploration time, subjects might be overwhelmed by the amount of information, try to keep up at the expense of exploring details, ask unnecessary questions or be more likely to abort the test.

Presenting mock-ups in binders can be problematic, when subjects are too curious. Some of our subjects were distracted by successive scenarios because they turned pages. An alternative could be to hand out individual sheets of paper. When doing so, the computer has to ensure that all mock-ups lie in the correct orientation in front of the subjects to avoid confusion. Therefore piles of mock-ups should not be circled around but rather be placed in front of each subject individually. Though, this procedure can be disruptive and time consuming. The facilitator should thus, when using binders, indicate during and in the beginning of the test when it is appropriate to explore which mock-up or when e. g. listening is desired.

Explaining. As blind subjects seldom have experience and knowledge about GUIs, it is essential to explain the design, content and purpose of the general design concept and the current mock-up from their perspective. This task can become a challenge, because difficulties in understanding might occur, even if the facilitator and his assistants are familiar with blind work techniques. It is advisable to take help from blind experts in advance and prepare explanations with them beforehand.

Additional illustration is also needed for specific tasks in applications that blind users are not familiar with, in order to allow them to comment on the implementation proposal. In our test, prior explanation of the mock-ups proved helpful because not all subjects possessed sufficient background knowledge.

4.3   Analysing

Shortly after the test, the results and impressions should be documented and analysed to avoid forgetting important details. The evaluation can comprise reading and completing the minutes, transcribing the audio to collect comments and evaluating the video. When video recordings are used, it is advisable to have ink printing superposed with embossing on one sheet. One could first print and then emboss or use a special printer which can do both simultaneously. In our test the sole embossed printings were hard to perceive on the recording when analysing the video, so that the elements explored in scenes could only be guessed by the positions of the subjects' hands.

After extracting impressions and comments of the subjects, it should be decided whether it is a problem of concept, material, explanation or a personal preference. When testing multiple mock-ups, e. g. for different scenarios, it is advisable to compare the comments concerning the different mock-ups to extract aspects which can be applied to the entire concept or design.

5   Recommendations

The main goal of an evaluation is to identify existing problems and to find potential for improvements. To achieve this aim, the evaluation must be prepared, conducted and analysed carefully and adequately. In the following, we present recommendations for tactile paper prototyping which have been condensed from our observations (see Section 4):

- Consider general issues when hosting blind users.
- Provide adequate facilities according to the special needs.
- Check for special hardware and software required.
- Design mock-ups not for sighted but for blind users.
- Even better, have blind people design the mock-ups.
- Proofread the mock-ups before conducting the test.
- Check the recording before and during the test, as repetition is expensive.
- Provide a sufficient number of assistants.
- Make sure that mock-ups are provided synchronously to each subject.
- Allow for sufficient time to explore the mock-ups.
- Explain from the blind users' perspective.

Different usability evaluation methods have different aspects which should be considered. Nevertheless, many of these recommendations do not only apply to tactile paper prototyping but are also applicable for other evaluation techniques involving blind subjects.

6   Conclusion and Outlook

Tactile paper prototyping is a new approach to design haptic user interfaces. It allows to bridge between the visual and haptic modality while ensuring multi-modality when using assistive technology to gain access to graphical user interfaces. Tactile paper prototyping applies to user centred design.

We developed 16 tactile paper mock-ups to design a user interface which includes Braille and tactile graphics using a planar tactile display. Due to limitations of resolution and size of such a tactile display re design of visual concepts is required. Like paper mock-ups, tactile mock-ups allow for verification of design concepts before implementation and involvement of end users. Therefore, we found it essential that blind people contribute to mock-up production and that mock-ups are evaluated by prospective blind users.

To aim at validation of our approach a follow-up study using a wizard-of-oz approach has been conducted. It involved three separate users who confirmed the suitability of a selection of haptic designs on the actual planar tactile device [14]. The mock-ups that were rejected during the tactile prototyping had deliberately not been included in this evaluation.

Future work will have to extend our approach to the design of audio-haptic interfaces possibly supporting also Braille-illiterate users. The suitability of tactile paper prototyping must be tested for other application areas such as maps, games, or collaborative software. The analysis of hand contact will have to be considered in more detail in order to understand failures and mismatches in the design more easily.

Acknowledgements. We thank all blind subjects who participated in our evaluations. We also thank Oliver Nadig and Ursula Weber for organising the tests and Christiane Taras for providing us with the software to create the mock-ups.

The HyperBraille project is sponsored by the Bundesministerium für Wirtschaft und Technologie (German Ministry of Economy and Technology) under the grant number 01MT07003 for Universität Potsdam and 01MT07004 for Technische Universität Dresden. Only the authors of this paper are responsible for its content.

References

1. Henry, L.S.: Just Ask: Integrating Accessibility Throughout Design (2007)
2. Völkel, T., Weber, G., Baumann, U.: Tactile Graphics Revised: The Novel BrailleDis 9000 Pin-Matrix Device with Multitouch Input. In: Miesenberger, K., Klaus, J., Zagler, W.L., Karshmer, A.I. (eds.) ICCHP 2008. LNCS, vol. 5105, pp. 835–842. Springer, Heidelberg (2008)
3. Rubin, J., Chisnell, D.: Handbook of Usability Testing: How to Plan, Design, and Conduct Effective Tests, 2nd edn. Wiley, Chichester (2008)
4. Schiewe, M., Köhlmann, W., Nadig, O., Weber, G.: What You Feel is What You Get: Mapping GUIs on Planar Tactile Displays. In: HCI International 2009 (2009)
5. Walker, M., Takayama, L., Landsey, J.: High-fidelity or Low-fidelity, Paper or Computer? In: Proceedings of the Human Factors and Ergonomics Society, pp. 661–665 (2002)
6. Snyder, C.: Paper Prototyping: The Fast and Easy Way to Design and Refine User Interfaces. Morgan Kaufmann, San Francisco (2003)
7. Rettig, M.: Prototyping for tiny fingers. Commun. ACM 37(4), 21–27 (1994)
8. Snyder, C.: Paper prototyping. IBM developerWorks (2001)
9. Magnusson, C., Brewster, S. (eds.): Guidelines for Haptic Lo-Fi Prototyping (2008)
10. Schwede, K.J., Klose, U.: EBSGO-Lehrerhandbuch: Grundlagen für den blinden- und sehbehindertenspezifischen Unterricht mit grafischen Oberflächen [EBSGO Teacher's Guide: Basics of Education for the Blind and Visually Impaired with GUIs]. Tectum-Verlag, Marburg (2001)
11. Tanhua-Piiroinen, E., Raisamo, R.: Tangible Models in Prototyping and Testing of Haptic Interfaces with Visually Impaired Children. In: NordiCHI 2008 (2008)
12. Taras, C., Ertl, T.: Interaction with Colored Graphical Representations on Braille Devices. In: HCI International 2009 (2009)
13. Dumas, J.S., Loring, B.A.: Moderating Usability Tests: Principles & Practices for Interacting, pp. 133–155. Morgan Kaufmann, San Francisco (2008)
14. Prescher, D.: Ein taktiles Fenstersystem mit Multitouch-Bedienung [A Tactile Window System with Multitouch Operation]. Diploma thesis, Dept. of Computer Science, Technical University of Dresden, p. 122 (2009)

The Carillon and Its Haptic Signature: Modeling the Changing Force-Feedback Constraints of a Musical Instrument for Haptic Display

Mark Havryliv[1], Florian Geiger[1], Matthias Guertler[1],
Fazel Naghdy[1], and Greg Schiemer[2]

[1] Faculty of Informatics
[2] Sonic Arts Research Network
University of Wollongong, NSW, Australia
mhavryliv@gmail.com, Florian.Geiger@iwb.tum.de,
{mrg059,fazel,schiemer}@uow.edu.au

Abstract. The carillon is one of the few instruments that elicits sophisticated haptic interaction from amateur and professional players alike. Like the piano keyboard, the velocity of a player's impact on each carillon key, or baton, affects the quality of the resultant tone; unlike the piano, each carillon baton returns a different force-feedback. Force-feedback varies widely from one baton to the next across the entire range of the instrument and with further idiosyncratic variation from one instrument to another. This makes the carillon an ideal candidate for haptic simulation. The application of synthesized force-feedback based on an analysis of forces operating in a typical carillon mechanism offers a blueprint for the design of an electronic practice clavier and with it the solution to a problem that has vexed carillonists for centuries, namely the inability to rehearse repertoire in private. This paper will focus on design and implementation of a haptic carillon clavier derived from an analysis of the Australian National Carillon in Canberra.

Keywords: Haptics, musical instrument, physical modeling, National Carillon.

1 Introduction

1.1 Haptics in Musical Instruments

It has been conclusively demonstrated that musicians and sound-makers depend heavily on their haptic interaction with a sound-producing device; it has also been demonstrated that performers are 'trainable', with an inherent capacity to learn new haptic cues and employ them in musical performance of novel instruments [1]. Indeed, novel instruments incorporating some form of haptics are increasingly pervasive as the barrier to entry for hardware and software is lowered [2].

However, sophisticated haptic implementations of traditional musical instruments are less common. The TouchBack piano [3], the V-Bow [4], the D'Groove [5], the MIKEY project [6], and the Haptic Drumstick [7] are notable examples of the few traditional instruments rendered specifically as haptic devices. Even rarer are attempts

M.E. Altinsoy, U. Jekosch, and S. Brewster (Eds.): HAID 2009, LNCS 5763, pp. 91–99, 2009.
© Springer-Verlag Berlin Heidelberg 2009

at applying haptic principles in realising instrument designs specifically designed to help train musicians in the performance of a traditional instrument.

In the authors' opinion, the greater research focus on haptics in novel, non-traditional devices is due to both legitimate interest in augmenting conventional instruments and creating new ones in order to extend the capabilities of electroacoustic performance, and the problems associated with recreating and simulating traditional instruments. These difficulties range from gathering information about the dynamic behavior of a traditional instrument to building a satisfactory prototype that has the 'feel' a seasoned instrumentalist expects.

A haptic incarnation of a traditional instrument, built for the purpose of practice or honing musicianship skills, must perform to the constraints of the real instrument. Further, a haptic instrument needs to replicate the visual, mechanical and sonic characteristics of the manipulandum – the point at which haptic interaction occurs between the musician and the instrument.

1.2 The Carillon Problem

The Haptic Carillon project is motivated by the possibility that the haptic and sonic characteristics of any carillon in the world can be simulated. To this end, an analysis of the haptic and sonic properties of the National Carillon in Canberra, Australia [8], has been undertaken, and these features have been statistically modeled.

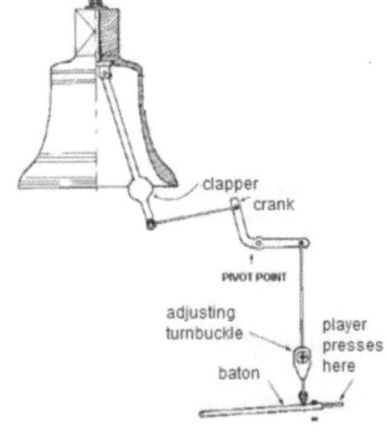

Fig. 1. (a) the carillon keyboard – National Carillon, Canberra, Australia (b) simplified representation of the carillon mechanism

A carillon (Figure 1) is a mechanical construction with bells of various size played by a carilloneur from a mechanical keyboard, or clavier, housed beneath the bell chamber. It represents a particularly difficult haptic/sonic problem.

The carillon exhibits a mechanical complexity comparable to the piano, but where the piano aims at a consistency of haptic response across the entire instrument the carillon is predicated upon the idea that each key requires a different degree of force to play. These forces vary widely across the instrument, and are subject to seasonal variation.

The sonic response of the carillon is also subject to change over time; while bell, and carillon, sound synthesis is an established research area [9, 10], it is also recorded that the sonic output of carillon bells change significantly over time [11]. What is less well-established is that the haptic behavior of a carillon also changes over time. This change is in no way linear or predictable, and it is not necessarily related to a change in the sonic behavior of the carillon.

The need for carilloneurs to develop musicianship and extend the instrument's repertoire offers a compelling musical reason to build a haptic practice instrument. Unlike other traditional instruments, the carillon, always has an audience, willing or unwilling, even if the carilloneur is only trying to practice.

We have developed the concept of a haptic signature as a way of acknowledging that a single type of instrument might have a variety of haptic behaviours, each of which is important to replicate if the instrument is itself to be haptically rendered.

2   The Carillon Mechanism

The National Carillon in Canberra, located in a tower on Aspen Island in Lake Burley Griffin, houses 55 bells spanning four and a half octaves. Each bell weighs between 7 kg and 6 ton. Despite the carillon's imposing mechanical construction its kinematic configuration is relatively straightforward.

Figure 1(b) is a simplified representation of the mechanism for one of the batons used to play the instrument. In its détente position, each baton rests against one of two beams that run horizontally across the range of the clavier, the upper beam for 'black' notes the lower for 'white' notes, in this position, the clapper on each bell is held away from the inside rim of the bell.

The bell clapper is connected to the baton via the bell crank. When a player presses downward on a baton, the clapper is pulled toward the inside of the bell. Between the upper and lower bells there is considerable variation in the force required to displace the clapper from its détente position. Measured at the tip of the baton this force is from 20-30 N for the lower bells to 1-3 N for the upper bells. This variation is continuous across the range of the clavier but is not linear; bell 4, for instance, requires 6N to displace the baton where bell 28 – at the halfway point in the keyboard – requires 1.3N. This component of the National Carillon haptic signature is shown in Figure 2.

2.1   The Haptic Signature

The overall trend can mostly be explained in terms of different clapper masses and lengths for different sized bells. However, differently configured springs in most baton mechanisms, and different initial angles at which the clapper is held (θ_i) significantly mitigate or exaggerate the differences in clapper mass.

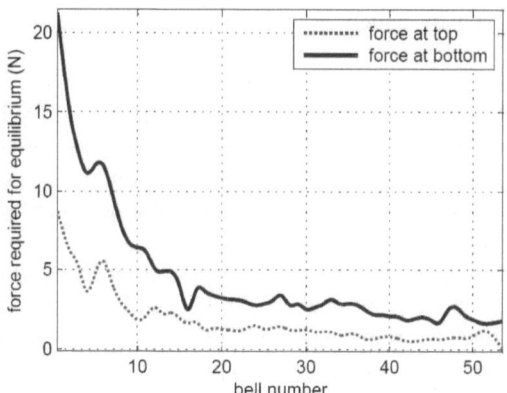

Fig. 2. Change in force required to bring the carillon system to static equilibrium across the range of batons in the National Carillon, Canberra. The baton's total displacement from the top of its stroke to the bottom is approximately -5cm.

The haptic signature above shows that the change in force required to play a baton is not linear, or even monotonic, across the range of the carillon. For the purposes of further modelling, we particularly note the difference between the force required at the top of the baton stroke and that required at the bottom of the stroke.

2.2 Dynamic Analysis and Modelling

For the purposes of analysis, the carillon model is divided into three rotational sub-models, each of which interacts with the other through reaction forces, otherwise interpreted as the tension in the two cables that join the baton system to the crank system (T_B), and the crank system to the clapper system (T_C). While these cables retain some tension (as they do throughout normal operation of the carillon), it is possible to define equations of motion for the respective rotational systems. A full description of the equations is presented in [12], however to facilitate further discussion, below are the three basic equations of motion for the respective rotational subsystems. Here, I is the sum of moments of inertia, θ is angular displacement, L is the full length and τ is the net torque occasioning from gravity acting on the components of each mechanism. The systems interact by solving for tension in the two cables.

$$I_B\ddot{\theta}_B + {}^{L_B}\!/_2\, T_B - \tau_B - F_P L_B = 0 \qquad (1)$$

$$I_{Crank}\ddot{\theta}_{Crank} - \tau_{Crank} + T_C L_{Crank} - T_B L_{Crank} = 0 \qquad (2)$$

$$I_{Clapper}\ddot{\theta}_{Clapper} - \tau_{Clapper} - b.\,\text{sign}(\dot{\theta}_{Clapper}).\,\theta_{Clapper}$$

$$- f\left(\dot{\theta}_{Clapper}\right) + T_C L_{Clapper} = 0 \qquad (3)$$

where

$$\tau_{Clapper} = g\sin\theta\,(\textstyle\sum_{i=1}^{3} m_i L_i) - L_{Clapper}(k\theta_{Clapper} + spring_i) \qquad (4)$$

$F_P L_B$ in (1) is the torque generated by the player applying force at the tip of the baton. $\{b.\,\text{sign}(\dot{\theta}_B)\theta_B\}$ in (3) models the position-dependant damping induced by the cable T_B – when the baton is fully depressed, the cable is pulled at an angle against a wooden frame lined with a felt-covering. $f(\dot{\theta})$ in (3) is a combined Coulomb (static) and viscous friction model. The three masses and lengths in (4) represent the three segments of the clapper, visible in Figure 1: the upper rod, the clapper ball, and the lower rod. As different bells are modelled, equations (1) and (2) are left unchanged; all variable parameters are localised in equation (3).

The variables in (3) can be grouped into those that generate torque from angular displacement and those that generate torque from angular velocity. Though possible, it is undesirable to model each bell by manipulating every variable independently. Two important simplifications are made: 1) curve-fitting techniques are used to estimate the lengths of the clapper components based on a limited set of measurements; and 2) post-facto analysis demonstrates that the change in torque over displacement can be well-fitted to simple linear polynomial equations.

2.3 Clapper Dimension Estimation

Each clapper mechanism is different from the next, both in the length of the entire mechanism, and the ratios between the positioning of the clapper ball and upper/lower rods. The physical inaccessibility of a clear majority of the bells in the carillon makes it very difficult to measure the dimensions of each clapper, so twelve clappers spaced across the carillon were measured and the change in lengths was plotted, then fitted using a smoothing spline function (Figure 3a).

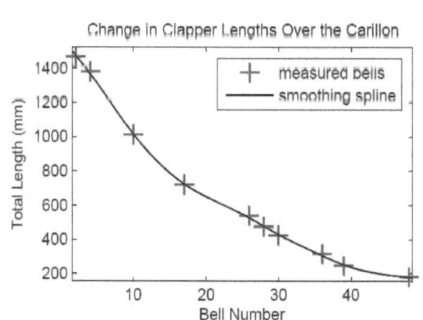 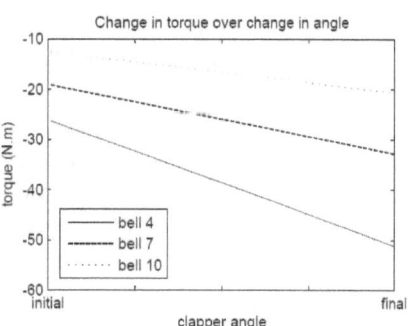

Fig. 3. (a) interpolated change in clapper mechanism lengths over the carillon (b) change in torque as a function of clapper position for bells 4, 7 and 10

Note the gradient of this graph is not particularly close to the changes in force shown in Figure 2, indicating that the torque generated by the clapper mass orientation is not the most important source of torque felt by the user. The other important determinants of the clapper's net torque are the clapper's initial angle, the force applied by the spring at the clapper's initial angle, the k value of that spring, and the friction induced by the motion of the cable T_B against the keyboard frame.

2.3.1 Change in Clapper Force – the Haptic Signature

While the crank and baton confer a mechanical advantage to the player that changes over the range of baton displacement, the primary sensation of force-feedback is provided by the changing clapper torque over baton displacement. Figure 3b shows this change in force for bells 4, 7, and 10, calculated simply by separating and solving for the position-dependant terms in equation (4).

The graph shows a near-linear change, and curve-fitting demonstrates it to be a linear polynomial function:

$$\tau[\theta] = k\theta + C \tag{5}$$

where k is the rate of change and C is an initial offset. The initial offset term encapsulates the mass and length components of the clapper mechanism, the clapper's initial angle, the initial force applied by the spring, and the maximum friction force generated against the cable T_B. The rate of change encapsulates the change in torque generated by the change in angle of the clapper mechanism, the spring k and the change in friction against the cable.

The linear function does induce a small amount of error due to the *sin* function in (4) effectively being eliminated, however this is intuitively negligible when recalling that, for small values of θ, $sin(\theta) \approx \theta$.

2.4 System Modelling

Applying the principle of uniform acceleration and calculating the relative accelerations between systems, it is possible to determine reaction forces at each cable connection and thoroughly model the dynamics of every bell in the National Carillon.

3 Models – Simulated and Measured Motions

The motion of the National Carillon was measured using an inertial sensor to compare the performance of the analytically derived model against the real carillon. All the mathematical models are programmed and executed step-wise in the Simulink ODE solver (www.mathworks.com/products/simulink).

3.1 Bell 4

Bell 4 is particularly well modelled, as it was reasonably accessible for the purposes of obtaining geometric and force measurements. The following figures demonstrate the performance of the simulated model against the measured data. Two initial tests are taken: releasing the baton from the bottom of its stroke in order to measure the dynamics of the system without user influence (Figure 4a), and releasing the baton from the top of its stoke with a mass attached in order to measure the system's response to applied force (Figure 4b).

It can be seen there is very little error between the simulated and measured data. This is supported by a static analysis of this bell in which forces recorded in Figure 2 are applied to the mathematical model and found to match the behaviour of the real carillon, i.e. finding equilibrium at the top and the bottom for the respective forces.

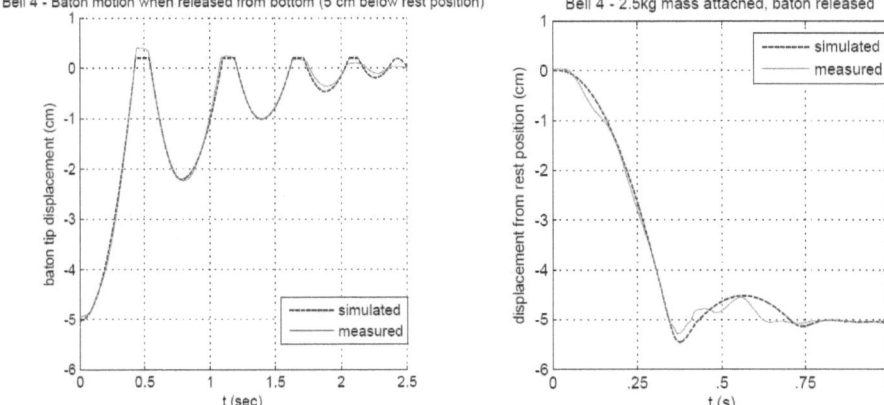

Fig. 4. Performance of the simulated model against measured data, a) for free motion of baton, and b) with 2.5kg mass attached to the tip of the baton

3.2 Bells 7 and 10

This method is proven for other bells; bells 7 & 10 modelled in Figure 5.

Note that the motions for bells 7 and 10 are quite different; in fact, bell 10 is a lot more similar in motion to bell 4 than it is to bell 7. The time is takes each baton to come to rest is a direct function of the difference in force required to stabilise the baton in different position as shown in Figure 2.

4 Prototype and Audio

The dynamic analysis presented in this paper is the basis for the haptic carillon prototype pictured below. The mathematical model is arranged such that it can be solved in real time using forward dynamics, i.e. the system's motion in response to forces. It is programmed in Simulink and compiled to run on a standalone target PC which connects to an electromagnetic linear actuator through a dedicated analog I/O board.

This actuator controls the position of the baton, and back-EMF at the actuator windings is measured in order to close the feedback loop by determining the force applied by the player.

Fortunately, the problem of generating appropriate sound synthesis is somewhat mitigated in this environment. Typically, a carillonneur only hears their instrument through loudspeakers amplifying the signal picked up by strategically-positioned microphones in the bell tower. The National Carillon, for example, provides only this type of aural feedback to the performer. The insulation of the playing room from the bell tower is very thorough.

The authors have recorded the carillon bells directly from these microphones, and by carefully mapping the velocity at which the baton was travelling when hitting the bell (propelled with a motor attached to a tachometer, and recorded at 15 velocities per bell) to the recorded sound, it has been relatively straightforward to design a playback mechanism in the real-time audio software environment, pure-data, that mixes

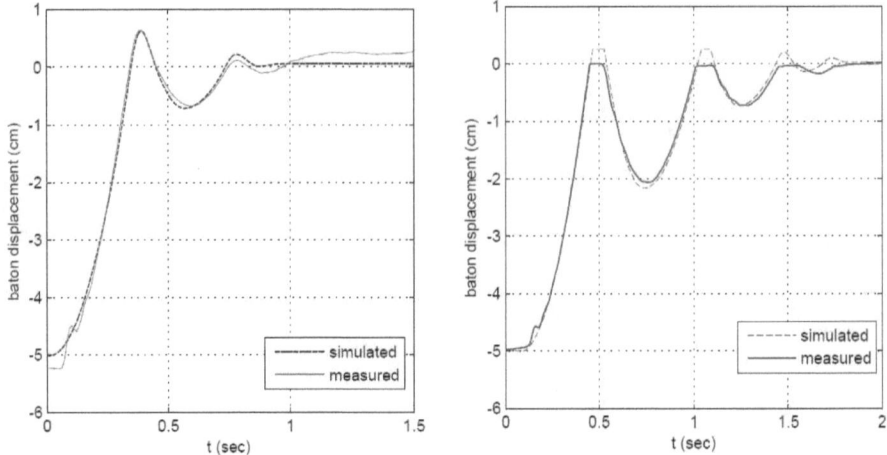

Fig. 5. Bells 7 and 10 modelled

Fig. 6. The haptic carillon single-baton prototype

the samples together based on the velocity with which a player has the clapper strike the bell. User-testing thus far indicates that carillonneurs are unable to distinguish between this method of sample playback and the amplification of an actual carillon bell signal in the tower.

5 Summary and Future Work

This paper has demonstrated analytical techniques that permit the accurate simulation of batons of varying force-feedback across the range of the National Carillon.

Future user-testing will build on current haptic research to assess the nature of a performer's perception of a traditional instrument against this haptically rendered one.

A particularly interesting avenue of enquiry will be researching the extent to which high-fidelity audio synthesis can mitigate low-fidelity haptic interaction – and vice versa – in the context of a replica of a traditional musical instrument.

References

1. O'Modhrain, S.: Playing by Feel: Incorporating Haptic Feedback into Computer-based Musical Instruments. Stanford University, Stanford (2000)
2. Berdahl, E., Steiner, H.C., Oldham, C.: Practical Hardware and Algorithms for Creating Haptic Musical Instruments. In: Proceedings of New Interfaces for Musical Expression, Genova, Italy (2008)
3. Gillespie, B.: Haptic Display of Systems with Changing Kinematic Constraints: The Virtual Piano Action. Stanford University, Stanford (1996)
4. Nichols, C.: The vBow: A Virtual Violin Bow Controller for Mapping Gesture to Synthesis with Haptic Feedback. Organised Sound Journal (2002)
5. Beamish, T., Maclean, K., Fels, S.: Designing the Haptic Turntable for Musical Control. In: Proceedings of 11th Symposium on Haptic Interfaces for Virtual Environment and Teleoperator Systems. IEEE, Los Alamitos (2004)
6. Oboe, R.: A Multi-Instrument, Force-Feedback Keyboard. Computer Music Journal 30(3) (2006)
7. Berdahl, E., Verplank, B., Smith III, J.O., Niemeyer, G.: A Physically-Intuitive Haptic Drumstick. In: Proceedings of International Computer Music Conference, Copenhagen, Denmark (2007)
8. National Carillon, Wikipedia,
   http://en.wikipedia.org/wiki/National_Carillon
9. Karjalainen, M., Esquef, P., Valimaki, M.: Making of a Computer Carillon. In: Proceedings of Stockholm Music Acoustics Conference, Stockholm (2003)
10. Fletcher, N.H., Rossing, T.D.: The physics of musical instruments. Springer, New York (1991)
11. Fletcher, N.H., McGee, W.T., Tanopolsky, A.Z.: Bell clapper impact dynamics and the voicing of a carillon. Acoustical Society of America 111(3), 1437–1444 (2002)
12. Havryliv, M., Naghdy, F., Schiemer, G., Hurd, T.: Haptic Carillon – Analysis and Design of the Carillon Mechanism. In: Proceedings of New Interfaces for Musical Expression, Pittsburgh, USA (2009)

Augmented Haptics – An Interactive Feedback System for Musicians

Tobias Grosshauser and Thomas Hermann

Ambient Intelligence Group
CITEC – Center of Excellence in Cognitive Interaction Technology
Bielefeld University, Bielefeld, Germany
{tgrossha,thermann}@techfak.uni-bielefeld.de

Abstract. This paper presents *integrated vibrotactiles*, a novel interface for movement and posture tuition that provides real-time feedback in a tactile form by means of interactive haptic feedback, thereby conveying information neither acoustically nor visually and it is a promising feedback means for movements in 3D-space. In this paper we demonstrate haptic augmentation for applications for musicians, since it (a) doesn't affect the visual sense, occupied by reading music and communication, (b) doesn't disturb in bang sensitive situations such as concerts, (c) allows to relate feedback information in the same tactile medium as the output of the musical instrument, so that an important feedback channel for musical instrument playing is extended and trained supportive. Even more, instructions from the teacher and the computer can be transmitted directly and unobtrusively in this channel. This paper presents a prototype system together with demonstrations of applications that support violinists during musical instrument learning.

Keywords: closed-loop tactile feedback, tuition, sensor, violin, bow, 3D-movement, real-time feedback.

1 Introduction

Musical instrument learning is a complex multi-modal real-time activity. It is representative for the larger class of human activity where expression and behavior shape and develop during practice towards a specific goal, similar to dance, etc. Due to its richness and complexity, novices tend to allocate their attention on the closed-loop interaction so that they comply with a coarse level of control, e. g. to produce the accurate frequency or to generate the accurate rhythm and this strong focus on primary objectives induces a neglecting of other important aspects such as a good body posture and alike that become relevant at later stages. Particularly, wrong coordination can even cause physical problems for musicians, and therefore techniques that can actively shift the player's focus of attention during practice are highly motivated.

In this paper we present an approach that uses tactile feedback as real-time feedback for the musician. Tactile feedback addresses our highly developed yet often neglected sense of touch. Extending or replacing visual or displays, tactile feedback is not limiting the focus of attention to attend a spatial location, it is wearable or

M.E. Altinsoy, U. Jekosch, and S. Brewster (Eds.): HAID 2009, LNCS 5763, pp. 100–108, 2009.
© Springer-Verlag Berlin Heidelberg 2009

connected with the used tools, here the violin bow and is highly capable to direct and alter the human's focus of attention directly at the right position, like a finger touch. Furthermore, we are capable to attend to even subtle cues in multi actuator scenarios simultaneously and perceive the vibrotactile stimuli as a whole at the same time.

With this motivation, the idea is to measure the player's motoric activity and to reflect specific properties of his/her performance as a task-specific and unobtrusive interactive haptic feedback, so that on the one hand, the musician can still focus on the musical sounds but receives additional information to keep awareness on relevant aspects of the physical execution. Practically, a prototype system has been developed and integrated into a common violin bow. The emitter is connected to an integrated chip, switching the small vibrotactile on and off and receiving control data via radio frequency transmission.

The vibrotactile feedback bow is our approach to integrate the essential technology in a tool that is typically in use anyway for musicians. It represents a first principled approach towards better closed-loop tactile interaction, here developed and optimized for a specific user group and application, but conceptually reaching beyond this case towards general vibrotactile based interaction support. The paper continues with a description of design aspects and a presentation of the technology. This is followed by a section on the selection of movements and requirements for the application of violin learning support. Finally we discuss our first experiences and our plans for continuation of this research.

2 Sensors and Actuators for the Closed-Loop Tactile Feedback System

This closed-loop tactile feedback, as shown in Fig. 1, supports musicians in learning, tuition, practicing and performing situations. Practicing and tuition are learning situations, if the output of a linear system (here the sensors) provides input for a learning system, the student or musician. This system is not only "learning", it is also "closed-loop", regarding the definition of Hugh Dubberly in [1], "if the learning system also supplies input to the linear system, closing the loop, then the learning system may gauge the effect of its actions and "learn" ".

In Sec. 3.2. we describe an exercise to be used to demonstrate the use of haptic feedback. It is an every-day situation, independent of age or skill level. Haptic feedback supports (a) the active learning phase by significant vibrotactile events for instruction in real-time, (b) the every-day practicing situations by long-term monitoring and increasing attention hints if fatigue symptoms appear, and (c) the non-solo performing situations by showing a new way and possibilities of musician-to-musician or conductor-to-musician communication. Besides these scenarios of violin learning, practicing and performing, other possibilities in the area of instrumental music or dance and sports are possible. The main parts of our prototype system are the acceleration and gyroscope sensors, the calculation of the feedback or receiving data via radio frequency transmission and the vibrotactile motors itself. The evaluation process includes sensor-based motion capturing, evaluated on music instrument learning and exercising based scenarios and closed-loop tactile feedback.

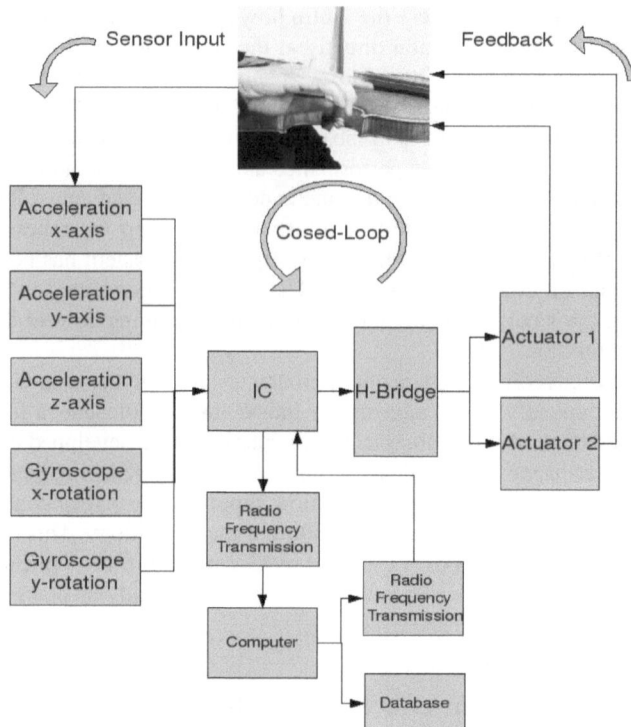

Fig. 1. The figure depicts the modular decomposition of our closed-loop system, showing the signal flow from sensing to actuators wrapped around the human in the loop.

2.1 Sensor Hardware for Sensing Motion

Similar to the carbon K-Bow form Keith McMillen [2], the "V-Bow" from Charles Nichols [3] and the used technologies in [4] and [5], acceleration and gyroscope sensor data were measured. In our exemplary use cases, 5 degrees of freedom, namely acceleration sensors for x-, y-, and z-axis and 2 gyroscopes are analysed. The data from the sensors are transmitted via radio frequency. A small Lithium Polymer (LiPo) battery is directly attached for power supply. The H-Bridge is an integrated electronic circuit, to apply a voltage to the vibration motors and changes the speed. Increased speed implies more urgency and attention of the musician, lower speed feels more soft. This small and light-weight sensor module can be used as a stand-alone tool, just for movement learning, or it can be clipped to a bow of a string instrument.

2.1.1 The Gyroscope

A IDG-300 dual-axis angular rate sensor from InvenSense is used. This allows the measurement of the rotation of the x- and y-axis of the bow stroke (see Fig. 2). The x-axis rotation is an additional compensating motion for e.g. soft bowing starts. The y-axis rotation is besides other functions relevant for pressure transfer onto the bow and to balance and change articulation and volume.

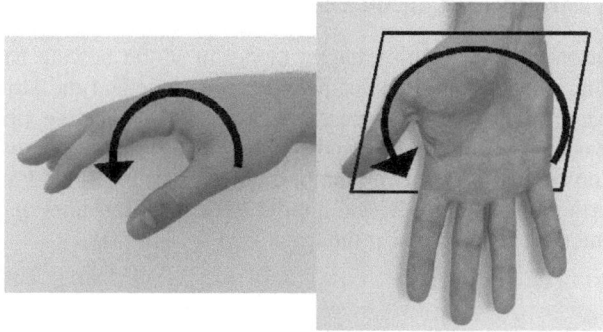

Fig. 2. The figure shows the x-axis (left) and y-axis (right) rotations of the player's hand, which corresponds to the sensed rotations using the bow-attached sensors during bowing activity

2.1.2 The Accelerometer

For the measurement of linear accelerations, the ADXL330 from InvenSense sensor is used, a small, thin, low power, complete x-, y-, and z-axis accelerometer. According to the description of the test cases in Sec. 3.1, every axis is important and has it's own defined plane, in which the movement is performed. Thinking in planes and rotations helps to learn complex movements, especially when the movement takes place beside your body and you can hardly see it or control it visually.

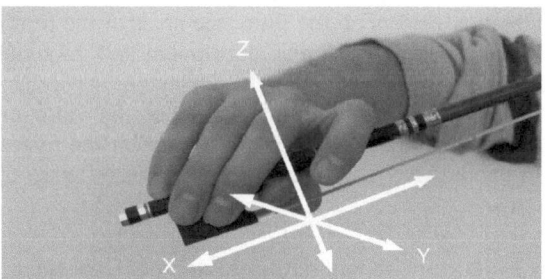

Fig. 3. Motion coordinate system, x-, y-, and z-axis

2.1.3 The Vibration Motor

Several vibration devices were taken into account, including simple vibration motors, solenoid piezo-electric elements and voice coils. Besides the simple control, weight and form factor, the availability and price have been important criteria for the choice. The vibration motor (see Fig. 4) with the dimensions 5x15mm, lightweight and cylindric shape seemed to be the best compromise. Furthermore, this kind of motor is typically used in mobile phones and is easy available for around 1€. Suitable vibration frequencies are around 250 Hz, since fingers and skin are most sensitive to these frequencies (see [6]).

2.2   Actuator Hardware for Tactile Feedback

In [7] a virtual environment with a haptic model of violin bowing has been implemented, focusing on the contact point between bow hair and the string and giving
physical feedback. In this paper we present a new developed active vibrotactile feedback system, easy to use, lightweight and attachable very flexible to manifold objects
such as violin bows. In this case two vibration motors (see Fig. 4 nr. 1 and 2) are fixed
to the bow. Furthermore a small IC and a radio frequency transmission module inclu-
sive a small battery is also fixed near the frog (see Fig. 4 nr. 3).

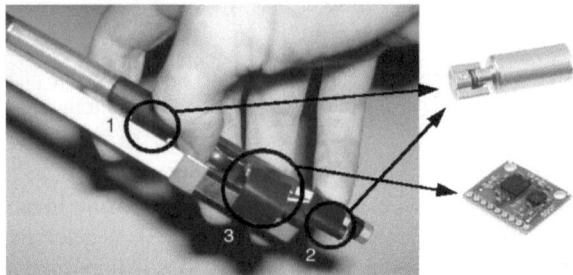

Fig. 4. 1 + 2=Vibration motor, 3=Sensor and battery

2.2.1   Description and Setup

The main goal was the mounting of the sensors, battery, radio frequency transmission
module and the vibration motors on the little free areas at the frog of the bow. They
are all fixed with adhesive foil for simple adjustment and removing. Also only one
vibration motor can be fixed, depending on the demand of the scenario.

The second important point was the placement of the vibration motors without
generating hearable vibrations or distortion. Fixation near the fingers came up to our
expectations of an unobtrusive, everyday usage without influencing the movements,
postures and gestures.

2.2.2   Data Flow Diagram with and without External Computer

The tactile feedback works with and without computer (see Fig. 5). In everyday scenarios and performance situations, the independence of computers is important. For
that reason, a simplified usage of sensors combined with data calculation on an IC is
possible. This system is cheaper and easier to use, because no radio frequency transmission is needed, and the radio module can be plugged off.

2.2.3   Listening with the Skin - Awareness of Tactile Feedback

The vibrations are short rhythmic bursts between 40Hz and 800Hz, which is the sensitive range of the mechanoreceptors in the fingers. The distance between the two motors is big enough for easy identification which one is vibrating. The amplitude and
frequency can be varied independently. This allows to evoke more or less attention,
increasing and decreasing of the vibration and at least 4 significant combinations between the two motors: (1) both motors on, (2) motor 1 on, motor 2 off, (3) motor 2 on

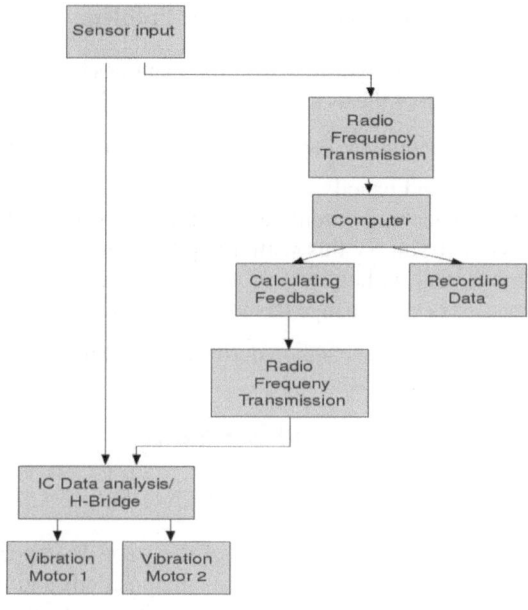

Fig. 5. Data flow diagram, with and without computer

and motor 1 off and (4) both motors off. As described in [8] the touch-sense feedback channel is extended and the awareness of the vibrotactile feedback is increased and trained.

3 Performance Aspects in Musical Instrument Learning

3.1 Methodical and Systematical Learning Scenarios for Complex Motor Skills

The following scenario is a basic extraction of beginners' violin lessons. Depending on the age of the pupil or student, different approaches exist. One of these is the breakdown and fragmentation of a movement into several simpler action units, based on the ideas of Conrad von der Goltz [9]. In our scenarios, a simple bow-stroke is decomposed in an exemplary use case. This is not only a beginner's problem, this is even trained from time to time by advanced students and professionals to develop their skills and physical awareness. The described sensor and the real-time tactile feedback gives us the possibility to train this simplified movement. The movement described in the following section can be performed simultaneously or successively, with or without instrument. The application of our system for the support of bowing movements is exemplary for many use cases in learning more or less complex movements. The capabilities of the closed-loop tactile feedback system cover different areas like sports, rehabilitation and other everyday scenarios.

3.2 Use Case of a Specific Motion Coordination Problem

Problem: Drawing a virtual straight line with the hand beside the body in the x- and y-planes with zero deviation of the x- and y-axis (see Fig. 3).

Pedagogical aspect: Understanding the "virtual straight line" of bowing movement.

Idea: If you move your hand exactly along one direction so that you draw a perfect line into the air beside your body, complex compensating movements of the hands and arms are necessary. If you try this with a pupil the first time, it is not only hard to understand the movement without seeing your hands, also practicing in front of a mirror is difficult, because every change has to be side-inverted.

Result: Students learn to move the hand on defined straight lines, without looking to it.

4   Interactive Tactile Design and Examples

4.1 Design of the Additional Haptic Feedback

In many teaching situations, body awareness of the pupil is important for gestures, postures and body movements. Tiredness causes malpositions and often a small verbal hint or touch on the concerned body part is enough to remember the correct position or movement. The closed-loop tactile feedback fits perfectly in this discreet way of indication and can support body awareness over a long period of time. Besides the mechanic feedback of the violin while playing the instrument and the uncomfortable feeling of a wrong body posture, the signals of the used mechanic vibrotactiles create passive touch cues, which are presented to the observer's skin, rather than felt in response to active movements, similar to [10]. As [11] states, "typically, this kind of functionality targets *multitasking environments* where the user's primary attention, as well as visual resources and possibly hands, are engaged in another task" .

4.2 Recommender System

For stroking the bow, the challenge is to move the bow along a straight line in 3D space. Such a task is enormously complex without any feedback. However, if the hand would be in contact with a wire under tension along the direction, the tactile feedback emitted to the hand would easily enable the hand to move along the line. The tactile feedback provides a guidance along which the hand can orient its movement. Obviously, feedback facilitates greatly the performance under such constraints. For the case of violin playing, a feedback of similar type can be implemented by triggering haptic feedback whenever a corridor of acceptable behavior is left. In doing so, the pupil can start with rather coarse constraints and gradually decrease the corridor size (with ongoing improvement) until ideal performance is obtained. The corridor of acceptable orientation can be defined by help of the sensor deviations (e.g. accelerometer or gyroscope) from an ideal orientation.

For instance, a tactile burst can be switched on whenever the orientation deviates more than a given threshold, for each direction in space via different tactile feedback frequency or tactile actuator.

The corridor could even be adjusted to be at 75% of the standard deviation in performance over the past 5 minutes. In doing so, both the progress is measured, and the system adapts to the performance of the pupil. It could also be quite motivating for a pupil to see such objective progress analysis over time. The described haptic feedback scheme is demonstrated in an interaction video on our website [12].

4.3   New Ways of Interaction with Silent Hints and Codes

In common one-to-one teaching situations, radio frequency based data transmission in combination with tactile feedback allows hints, physically executed with the vibration motor, controlled by the teacher instead of disturbing verbal reminders. The problem in teaching and learning situations is, that after a while or a short played phrase, problems discussed shortly before are forgotten. But mostly a small hint e. g. from the teacher, is enough, to remember it. Usually this is a spoken word, but here, a short physical hint is used. But if you define a short "reminder-sequence", e. g. three short vibration impulses from motor 1, motor 2 and motor 1 again, while discussing the problem before the performance, it makes the remembering during the performance for both sides, the pupil and the teacher, who executes the sequence, "the physical hint" with the vibrotactile in the right moment, much easier. This is just one of many examples, how such hidden and silent hints can be used.

5   Discussion

The possibilities of the sensor data based closed-loop tactile feedback system will show in an intuitive way, that every change of the movement, here in 3D-space can be signaled unobtrusively. Even a real-time correction or an overdone correction can be shown. The signals help to understand quite intuitively, how a special movement, in this case the bowing on a stringed instrument, works. To our first experience, it supports the learning effect and the optimization of the movement, only by the sense of touch. With "Hearing with the skin", the very underestimated and often neglected but important feedback channel for musicians, the vibrotactile feedback will be furthered and trained like the active listening ability, not only by music students and pupils.

Our first impression is that the continuous sensor data based closed-loop tactile feedback described above works well and is quite efficient to direct the attention to improper executions. As promising prospect, the system may lead to learning aids for visually impaired people, especially as they are more biased to use their non-visual senses to compensate the lacking visual information.

6   Conclusion

This paper has introduced the sensor data based closed-loop tactile feedback system as a portable, integrated, versatile, interactive system for musicians. The system combines sensor technology, real-time tactile feedback, and new ideas on learning movements, postures and gestures into a usable every day system. The presented application has been specifically selected and optimized for the task of violin learning and the closed-loop tactile feedback demonstrates that the vibrotactile feedback and impact sound conveys useful information. We plan to conduct long-term user studies with this proto-type and we'll find more scenarios of competitive and useful closed-loop tactile feed-back. We hope that our closed-loop tactile feedback system can make a positive contribution to better pedagogic approaches and methodical understanding, exercising-productivity rising methods and ultimately to the development of more healthy prac-tices for musicians.

Finally, we are very convinced that we can easily adapt the system to other musical instrument playing problems and even to other fields such as movement training in sports and dance, where the closed-loop tactile feedback system can help to better learn, understand and perform complex movements.

Acknowledgement

The work has been supported by CITEC – Center of Excellence in Cognitive Interac-tion Technology.

References

1. Dubberly, H., et al.: What is interaction? Are there different types? Interactions 16(1), 69–75 (2009)
2. McMillen, K.: Stage-Worthy Sensor Bows for Stringed Instruments. In: NIME 2008, Casa Paganini, Genova, Italy (2008)
3. Nichols, C.: The vBow: Two Versions of a Virtual Violin Bow Controller. In: Proceedings of the International Symposium of Musical Acoustics, Perugia (2001)
4. Bevilacqua, F.: The augmented violin project: research, composition and performance report. In: NIME 2006, Paris, France, pp. 402–406 (2006)
5. Young, D.: Wireless sensor system for measurment of violin bowing parameters. In: Music Acoustics Conference, Stockholm, Sweden, pp. 111–114 (2003)
6. Marshall, M.T., Wanderley, M.M.: Vibrotactile Feedback in Digital Musical Instruments. In: NIME 2006, Paris, France, pp. 226–229 (2006)
7. Baillie, S., Brewster, S.: Motion Space Reduction in a Haptic Model of Violin and Viola Bowing. In: Eurohaptics Conference, 2005 and Symposium on Haptic Interfaces for Vir-tual Environment and Teleoperator Systems, 2005. World Haptics 2005, First Joint, vol. 18-20, pp. 525–526 (2005)
8. Bird, J., Holland, S.: Feel the Force: Using Tactile Technologies to Investigate the Ex-tended Mind. In: Proceedings of Devices that Alter Perception, DAP 2008, Seoul, North Korea (2008)
9. Goltz, v.d.C.: Orientierungsmodelle fur den Instrumentalunterricht. G. Bosse Verlag, Regensburg (1974)
10. Gibson, J.J.: Observations on active touch. Psychol. Rev. 69(6), 477–490 (1962)
11. MacLean, K.E., Hayward, V.: Do It Yourself Haptics: Part II. Robotics & Automation Magazine 15(1), 104–119 (2008)
12. Großhauser, T., Hermann, T.: Online demonstrations for publications (2009),
    http://www.sonification.de/publications/
    GrosshauserHermann2009-AHA

Interaction Design: The Mobile Percussionist

Tiago Reis, Luís Carriço, and Carlos Duarte

LaSIGE, Faculdade de Ciências, Universidade de Lisboa
treis@lasige.di.fc.ul.pt, {lmc,cad}@di.fc.ul.pt

Abstract. This paper presents the user centered iterative interaction design of a mobile music application. The application enables multiple users to use one or more accelerometers in order to simulate the interaction with real percussion instruments (drums, congas, and maracas). The ways through which the accelerometers are held, before and during interaction, define the instruments they represent, allowing the swapping of instruments during musical performances. The early evaluation sessions directed to the interaction modes created for each instrument enabled design iterations that were of utmost importance regarding the final application's ease of use and similarity to reality. The final evaluation of the application involved 4 percussionists that considered it well conceived, similar to the real instruments, natural and suitable for entertainment purposes, but not for professional musical purposes.

Keywords: Audio Interaction, Context Awareness, Mobile Interaction, Accelerometer-Based Gesture Recognition.

1 Introduction

Mobile music is a new field concerned with musical interaction in mobile settings through the utilization of portable technology [1], such as laptops, tablet PCs, mobile phones, and PDAs, amongst others. These devices have long been recognized as having potential for musical expression [2]. Such potential enables entertainment and artistic applications to leave the stationary living-room set, detaching themselves from the graphical and audio output of televisions (e.g. Wii), therefore, becoming available anywhere, allowing and encouraging multiple users to engage in spontaneous jam-sessions.

On another strand, the availability of an immense variety of sensors, their low prices, and their recent and rising inclusion on mobile devices enables the creation of mobile sensor-based musical applications. As we show in this paper, and as others have shown in previous experiments and products [4] (e.g. Wii), the use of sensor technology brings simulated musical interaction closer to reality [6]. It enables and facilitates the creation of input modalities that replicate musical interaction with different, well-known, real musical instruments (or families of musical instruments).

The research presented on this paper focuses the user centered iterative design of a multi-user mobile music application: The Mobile Percussionist. This application uses movement and grasp, both inferred from acceleration, as input, triggering percussion audio samples as output. It simulates three well-known percussion instruments, which

M.E. Altinsoy, U. Jekosch, and S. Brewster (Eds.): HAID 2009, LNCS 5763, pp. 109–118, 2009.
© Springer-Verlag Berlin Heidelberg 2009

are played in different ways: drums (drumstick strikes), congas (hand strikes), and maracas (shaking movements); attempting to imitate, as much as possible, the interaction with these instruments' real counterparts. The application makes use of one or more accelerometers, wirelessly connected to a mobile device. These accelerometers are used in order to simulate the interaction with real percussion instruments, and the way through which they are held defines the instruments they represent, introducing context awareness principles in order to enable the intelligent selection and swapping of the different percussion instruments that are simulated (drums, congas, and maracas). The differences between how these instruments are played presented a motivating challenge for the interaction design of this application regarding both the selection of the instrument simulated by each accelerometer and the interaction modes created for each instrument. As it was being developed, the application was evaluated through strongly user-centered procedures, presenting very good results for entertainment purposes.

The following section presents the work developed in the mobile music field, emphasizing the differences with the research presented on this paper. Afterwards, we present and discuss the iterative user centered interaction design of the Mobile Percussionist. Subsequently, we describe and discuss the final evaluation of the different interaction modes available on the application, and, finally, we conclude and present future work directions for the Mobile Percussionist.

2 Related Work

The field of mobile music is developing fast over the past few years. A clear fact supporting this affirmation is the recent creation of the Mobile Phone Orchestra (MoPhO) [3] of the Center for Computer Research in Music and Acoustics of Stanford's University. MoPhO is the first repertoire and ensemble based mobile phone performance group of its kind and has already performed concerts. The ensemble demonstrates that mobile phone orchestras are interesting technological and artistic platforms for electronic music composition and performance.

The creation of the abovementioned orchestra was only possible because researchers have developed different mobile music applications, some of which are becoming very famous and broadly used. One good example is the Smule OCARINA [4], which is sensitive to blow, touch and movements. This application does not use precompiled riffs and allows users to select between diatonic, minor and harmonic scales. Moreover, it is a social application. Users can see and hear other Ocarina players throughout the world and rate their favorite performances, enabling other users to benefit from their judgments.

CaMus [5] - collaborative music performance with mobile camera phones - demonstrates that mobile phones can be used as an actively oriented handheld musical performance device. To achieve this, the visual tracking system of a camera phone is used. Motion in the plane, relative to movable targets, rotation and distance to the plane can be used to drive MIDI enabled sound generation software or hardware.

Closer to our goals, ShaMus [6] - a sensor-based integrated mobile phone instrument - presents a sensor-based approach to turn mobile devices into musical instruments. The idea is to have a mobile phone be an independent instrument. The sensors used are accelerometers and magnetometers. The sound generation is embedded on the phone itself and allows individual gestural performance through striking, shaking and sweeping gestures. The striking gestures presented enable users to play at most 4 drum components (base, snare, tom, and hi-hat). To prove this concept a Nokia 5500 was used, connected to a Shake unit [7], which is a small device that incorporates a range of high-fidelity sensors for rapid prototyping of mobile interactions. The Shake's core unit contains a 3-axis accelerometer, a 3-axis magnetometer, a vibration motor for vibro-tactile display, a navigation switch, and capacitive sensing abilities.

The Mobile Percussionist is significantly different from ShaMus. We do not limit the number of accelerometers used. This introduces the possibility of having multiple users playing music at the same time, while using only one mobile device with sound processing capabilities. Moreover, our approach enables users to perform natural actions that simulate the interaction with different percussion instruments, allowing them to swap instruments while playing. Finally, the interaction modes of the Mobile Percussionist were iteratively designed through strongly user centered procedures where accuracy rates for the different actions were registered. These procedures also considered users' opinions regarding the actions' ease of use and naturalness when compared to real percussion instruments.

3   Iterative User Centered Interaction Design

This section describes the user centered interaction design of the Mobile Percussionist. The process focused the creation and evaluation of different interaction modes for each one of the three instruments considered, as well as the creation and evaluation of a mechanism that enables the application to be continuously aware of which instruments are being simulated by each accelerometer in use.

Accordingly, the process comprised the development and evaluation of different prototypes. The prototypes created were developed using NetBeans IDE 6.1, Java 1.6.0_07, two SunSPOTS [8], one laptop, and a set of stereo speakers. SunSPOTS (Fig. 1) are small devices with processing capabilities. These devices include two buttons and a variety of sensors: one 3 axis accelerometer, one light sensor, and one temperature sensor. Furthermore, they can also work as a platform where other sensors can be connected. The mentioned accelerometer has a sensitivity of either 600 mV/G or 200 mV/G relative to the selected scale +/-2G or +/- 6G, respectively [8]. In our experiments we used the 6G scale.

The early evaluation sessions, which guided the iterative user centered design of this application, focused the different prototypes created and considered all the actions supported by each interaction mode available in the application. The main goal was to understand the accuracy rates for each action and users' opinions about the similarities between the interaction modes created and the interaction with real instruments. Four users were involved, all male, with ages comprehended between 20 and 30 years old. Two of these users had previous experience with percussion instruments while the other two had experience with simulated percussion instruments (e.g. Wii Drums).

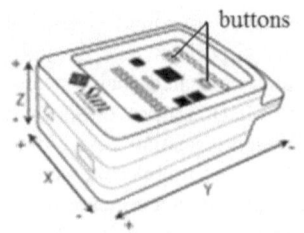

Fig. 1. SunSPOT and Accelerometer Axis

Following we describe the iterative user centered design of the interaction modes developed for each instrument considered, and afterwards, the design of the instrument selection and swapping algorithm.

3.1 Drums Interaction

Initially, in order to create an interaction mode that simulates real drums interaction, the design team focused only on accelerometer data. We defined the way the devices should be held, attempting to simulate drumsticks' grasp (notice that the leds are facing up in Fig. 2 and Fig 5). Following, it was necessary to accurately identify natural gestures on the accelerometer data. Three actions were defined, visually identified (inverted acceleration peaks), and labeled (Fig. 2): X-Strikes (fast movements performed sideways◐); Y-Strikes (fast movements performed forwards and backwards●); and Z-Strikes (fast movements performed up and down ○).

In order to programmatically identify these actions, a simple algorithm was implemented and evaluated in terms of accuracy. The actions were evaluated in terms of ease of use and similarity to reality. Each user performed each available action 20 times, in a random order with the purpose of reducing bias. Average accuracy rates

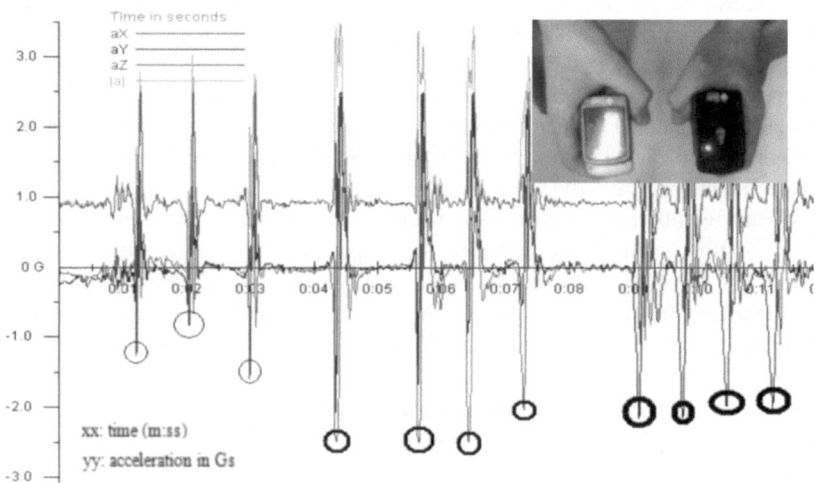

Fig. 2. Accelerometer Data: Strike Gestures (from the left to the right: Z, X, Y)

were of 90% for X-Strikes, 85% for Y-Strikes, and 95% for Z-Strikes. All users reported that performing Y-Strikes was very unnatural, and, therefore, more difficult than the remainder actions. Moreover, they all emphasized that Z-Strikes were the most natural ones regarding the interaction with drums. This early evaluation procedure was of utmost importance, since it revealed that only 2 of the 3 gestures were considered natural enough and that one of those was more natural than the other.

Afterwards, our team faced another design issue. Drums are composed by many components (e.g. snares, toms, crash, hi-hat, etc.), and when users are interacting with them they can "jump" directly from one component to another. Accordingly, the two gestures, which were previously considered natural enough for the interaction with drums, were not sufficient to simulate such a complex interaction. Therefore, in order to increase the number of samples available on each accelerometer, we considered the use of the 2 buttons available on each SunSPOT. The use of such buttons enables the definition of 4 different states. When these states are combined with the 2 previously elected strike actions, 8 different samples become available on each accelerometer. However, two questions still remained, whether to consider both strike actions or only one, and whether to consider combinations of buttons or not? In order to answer these questions, two prototypes were created and evaluated. Both prototypes considered the use of the 2 buttons available on the SunSPOTS. These buttons were used in order to select the audio samples, and the strikes to trigger those samples. The difference between the prototypes relied on the triggering of the samples. One considered only Z-Strikes and the other considered both Z and X-Strikes.

The same four users evaluated each prototype created. All of them reported that the use of Z-Strikes created an interaction mode that was more similar to reality. However, all the users agreed that the use of X-Strikes, despite being less natural, should also be considered, since it doubles the number of samples that can be triggered by each accelerometer. Finally, all users reported difficulties while having to press more than one button at a time, suggesting that the use of button combinations in order to increase the number of samples available is not a good approach.

Accordingly, the drums interaction mode available on the Mobile Percussionist considers the use of all the buttons (per se) available on the Sunspots, Z-Strikes, and X-Strikes, allowing the use of 6 audio samples per accelerometer.

3.2 Congas Interaction

The design of the congas interaction mode started by considering only accelerometer data. Similarly to the interaction design presented above, the main goal was to create an interaction mode that simulated as much as possible the interaction with real congas. We started by defining the way the devices should be held, in order for the application to be aware that the user is playing congas (notice that the leds are facing down in Fig. 3 and Fig. 5).

Following, we decided to consider 2 different actions that are used when playing real congas: center strikes (Fig. 3 on the left) and edge strikes (Fig. 3 on the right). Again, these actions were visually identified, and, afterwards, programmatically identified. Their distinction is inferred from the average acceleration on the Y-axis (O). When holding the device in the two different ways shown in Fig. 3 ways, the average acceleration on the Y-axis varies significantly enough to enable the application to be

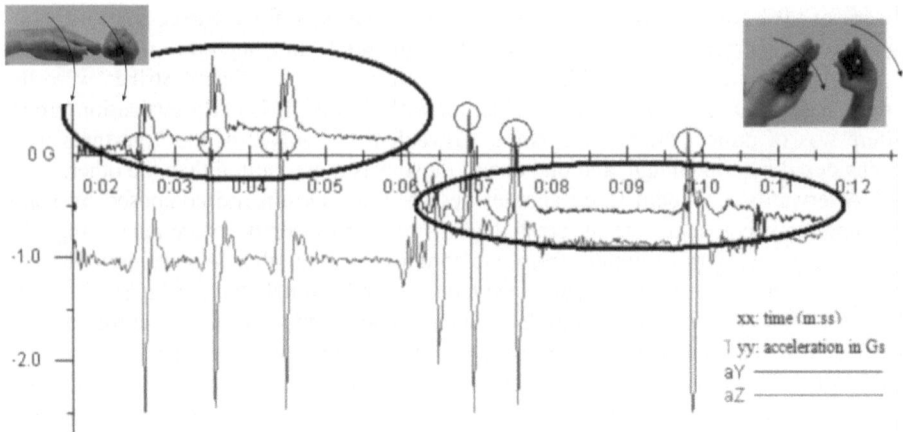

Fig. 3. Playing Congas

aware of users' intentions regarding the type of strike they are performing: center strikes consider a positive average acceleration, while edge strikes consider a negative average acceleration. Finally, Z-Strikes trigger the samples. Nonetheless, contrarily to the drums interaction mode, and since the way the accelerometers are being held is different, the Z-Strikes are represented by acceleration's peaks (○) instead of inverted peaks.

The users mentioned on the introduction of this section were all involved on the evaluation of this interaction mode. Again, each user performed each action 20 times, in a random order with the purpose of reducing bias. The procedure revealed average accuracy rates of 90% for both center strikes and edge strikes. Users reported that this interaction mode was very similar to the real interaction with congas. However, they disliked the limitation of using just one conga, since, usually, percussionists use more than one.

Similarly to what was done with the drums interaction, we decided to use the buttons available on the SunSPOTS, in order to increase the number of congas available. Two prototypes were created and evaluated by the same users. The difference between these prototypes relied on the way the buttons are used. One prototype considered holding the buttons pressed as state activators, while the other would activate and deactivate states when the users pushed the buttons.

The results were very similar for both prototypes. Users were more satisfied with the increased number of samples available, but all of them reported that interaction became a little less natural than before. Moreover, all the users agreed that holding the buttons pressed implied less cognitive effort than pushing the buttons in order to activate states, since the latter demands the memorization of the actual state in use.

Accordingly, the congas interaction mode available on the Mobile Percussionist considers the use of all the buttons (held pressed and per se) available on the Sun-SPOTS, Center-Strikes, and Edge-Strikes, allowing the use of 6 audio samples per accelerometer (3 different congas).

3.3 Maracas Interaction

The design of the maracas interaction mode considered only accelerometer data. As for the previous instruments, the main objective was to simulate reality as much as possible. We started by defining the way the devices should be held, in order for the application to be aware that the user is playing maracas (notice that the device is in a vertical position in Fig. 4 and Fig. 5).

Accordingly, two audio samples can be triggered by each accelerometer. One relates to shaking the maraca and the other transmits the sound of stopping the shaking. This interaction mode joins principles of the two abovementioned designs. Both peaks (○) and inverted peaks (**O**) are considered, the first triggering one audio sample and the latter the other.

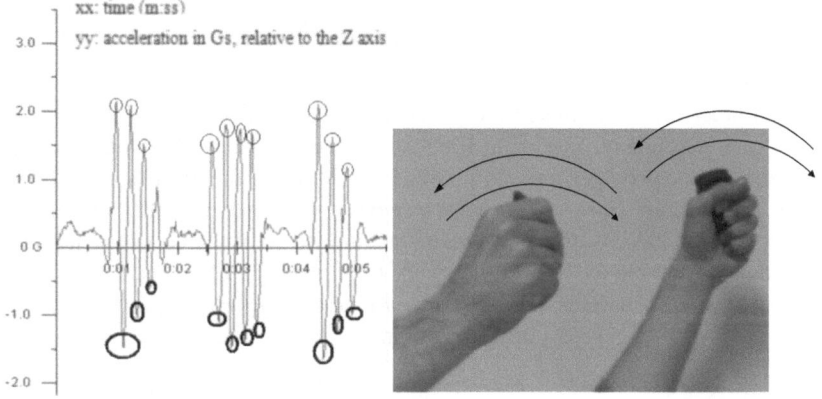

Fig. 4. Playing Maracas

The four users, mentioned on the introduction of this section, performed each available action 20 times. Average accuracy rates were of 95% for both actions. All users reported that this interaction mode was very similar to the real interaction with maracas and, therefore, very natural.

3.4 Instrument Selection and Swapping

Once the application is initiated, the users select the instrument they want to associate to each accelerometer in use. This selection (Fig. 5) is based on the way the devices are held, which is inferred from the acceleration on the Z-axis, and activates/deactivates the different interaction modes available. While playing an instrument, the acceleration on the Z-axis varies considerably. However, the average-based analysis behind the instrument selection and swapping mechanism enables the identification of the instrument being played also during interaction, allowing users to swap between instruments, depending on the way they hold the accelerometers.

The evaluation procedure considered the initial selection of the instruments and the swapping between instruments while playing. Again the order of the experiments was

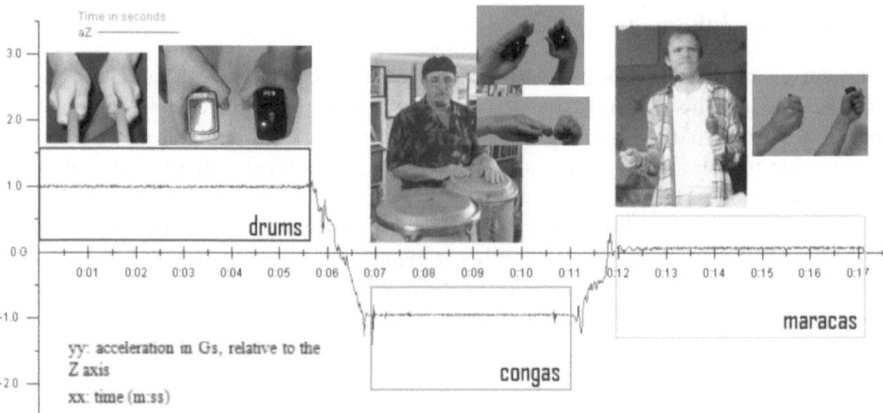

Fig. 5. Initial Step: Instrument Selection

defined in order to reduce bias. Firstly, each user involved performed 5 initial selections for each instrument, averaging 100% of accuracy. Afterwards, while playing the instruments, each user performed 5 swaps for each of the 6 possible transitions between instruments. Here, the average accuracy was also of 100%. However, the response time on the swapping between instruments caused the triggering of one unwanted audio sample on 30% of the swaps (e.g. playing a drum audio sample while the user was already holding the device as a maraca). Nevertheless, these unwanted samples were only triggered on the beginning of the experiment, suggesting a very short learning curve.

3.5 Discussion

The primary goal of this design process was to create natural interaction modes that simulate the interaction with real percussion instruments, allowing the swapping of instruments during musical performances. This goal could only be achieved through the iterations of the user centered design process conducted.

The first prototypes designed considered only accelerometer data, which was enough to enable the natural swapping of instruments. However, the experiments abovementioned, especially the ones regarding the interaction with drums, led us to use the buttons available on the devices, in order to increase the number of audio samples triggered by natural gestures. These gestures were elected by the users involved and all the possible combinations of buttons were experimented. The use of more than one button at a time decreased interaction's naturalness significantly, although the use of the each button per se did not. The use of the buttons was also included on the interaction with congas, but not on the interaction with maracas.

At the end of the process the users involved were satisfied with the different interaction modes created, their ease of use, and with the possibility of swapping instruments while playing.

4 Final Interaction Evaluation

This section is dedicated to the final qualitative evaluation of the Mobile Percussionist. The procedure involved four percussionists, none of them involved on the previous evaluation sessions, all male, with ages comprehended between 18 and 36 years old. There was a special concern in understanding if the users were capable of successfully playing rhythms with the application and make use of all the different instruments that are available. Accordingly, the users involved were taught to interact with the application through its different features, and were encouraged to use it freely to its full extent during approximately 20 minutes. Finally, the users expressed their difficulties and opinions during an interview.

All the users involved considered the concept of the application very interesting and useful for entertainment purposes, but not for professional musical purposes. They underlined the possibility of swapping between instruments while playing as a very innovative feature, emphasizing also the similarities between the interaction modes of the simulated instruments and their real counterparts as a very intelligent, usable, and natural feature, especially the ways the accelerometers are held.

Concerning the interaction with drums, all the users considered that six audio samples per accelerometer was a small number to simulate a drum set to its full extent. The same issue was reported regarding the interaction with congas, however, in this case, only two of the users considered three congas per accelerometer as a small number. Regarding the interaction with maracas, all the users considered that the number of samples was appropriate, however, all of them expressed that the buttons could be used in the same way they are for the congas and drums, enabling the use of three types of maracas instead of one.

Another issue, reported by all the users involved, consisted on the lack of expressiveness of all the interaction modes designed. This happened because our application makes use of accelerometer data only in order to trigger the samples used and not to define the strength with which these are performed (velocity). However, this data could also be used for this purpose.

Finally, all the users reported minor issues related to their own expressiveness while playing. All these users were percussionists, and, consequently, they had previous experience playing the real instruments our application tries to simulate. Accordingly, these users were experiencing difficulties during the first two/three minutes of their free play because their expressiveness caused an unwanted instrument swapping. However, after realizing this, these users voluntarily reduced their expressiveness being able to play fairly elaborated rhythms composed by the three available instruments.

5 Conclusion and Future Work

The research presented on this paper focused the iterative user centered interaction design of a mobile accelerometer-based percussion application. This application uses movement and buttons as input mechanisms, triggering percussion audio samples as output. It introduces context awareness principles that enable the swapping between the different percussion instruments available (drums, congas, and maracas) during musical performances. The different interaction modes available were designed and developed through iterative user centered procedures that significantly increased their similarities to reality. All the interaction aspects regarding the use and swapping of instruments were evaluated by four percussionists that considered that the application was innovative and simulated reality well enough to be used for entertainment purposes, but not for professional musical purposes.

Our future work plans, regarding the Mobile Percussionist, include porting the application to mobile phones, increasing the application's mobility factor and enabling the use of built-in accelerometers whenever those exist on the host device. However, we want to keep the possibility of connecting multiple accelerometers via wireless technology (e.g. SunSPOTS through Bluetooth), maintaining the multi-user facet of the application. Once this port is complete we intend to evaluate the application in order to understand discrepancies between its performance on a laptop and on a mobile phone. Moreover, we intend to test the application, both in the laptop and mobile phone versions, in order to understand how many accelerometers can be connected at the same time without generating or increasing latency significantly enough for it to be heard. We intend to use MIDI synthesizers in order to increase the expressiveness of the interaction modes available. Finally, we envision the inclusion of haptic feedback through the connection of vibrating devices to the SunSPOTS on the laptop version of the application, and through the use of mobile phones' vibrators on the ported version of the application.

Acknowledgments. This work was supported by LaSIGE and FCT through the Multiannual Funding Programme and individual scholarships SFRH/BD/44433/2008.

References

1. Gaye, L., Holmquist, L.E., Behrendt, F., Tanaka, A.: Mobile Music Technology: Report on an Emerging Field. In: Report - NIME 2006, Paris, France (2006)
2. Essl, G., Wang, G., Rohs, M.: Developments and Challenges Turning Mobile Phones into Generic Music Performance Platforms. In: Proceedings of Mobile Music Workshop, Vienna (2008)
3. Wang, G., Essl, G., Pentinnen, H.: MoPhO: Do Mobile Phones Dreams of Electric Orchestras? In: Proceedings of the International Computer Music Conference, Belfast (2008)
4. Smule OCARINA, http://ocarina.smule.com/
5. Rohs, M., Essl, G., Roth, M.: CaMus: Live Music Performance using Camera Phones and Visual Grid Tracking. In: Proceedings of the International Conference on New Interfaces for Musical Expression (2006)
6. Essl, G., Rohs, M.: ShaMus - A Sensor-Based Integrated Mobile Phone Instrument. In: Proceedings of the International Computer Music Conference (ICMC), Copenhagen, Denmark, August 27-31 (2007)
7. Hughes, S.: Shake – Sensing Hardware Accessory for Kinaesthetic Expression Model SK6. In: SAMH Engineering Services, Blackrock, Ireland (2006)
8. Sun Sunspots, http://www.sunspotworld.com

Vibratory and Acoustical Factors in Multimodal Reproduction of Concert DVDs

Sebastian Merchel and M. Ercan Altinsoy

Chair of Communication Acoustics, Dresden University of Technology, Germany
sebastian.merchel@tu-dresden.de

Abstract. Sound and vibration perception are always coupled in live music experience. Just think of a rock concert or hearing (and feeling) a church organ sitting on a wooden pew. Even in classical concerts kettle-drum and double bass are sensed not only with our ears. The air-borne sound causes seat vibrations or excites the skin surface directly. For some instruments (e.g. an organ) structure-borne sound is transmitted directly from the instrument to the listener.

If concert recordings are played back with multimedia hi-fi systems at home, these vibratory information is missing in the majority of cases. This is due to low reproduction levels or to the limited frequency range of conventional loudspeakers. The audio signal on todays DVDs contains an additional channel for low frequency effects (LFE), which is intended for reproduction using a subwoofer. The generation of tactile components is still very restricted. An enhancement of such a systems might be possible using an electrodynamical shaker which generates whole body vibration (WBV) for a seated person.

This paper describes a system implementing this approach. The generation of a vibrotactile signal from the existing audio channels is analyzed. Different parameters during this process (amplitude of the vibration, frequency range) are examined in relation to their perceptual consequences using psychophysical experiments.

Keywords: Multimodal Music Reproduction, Whole Body Vibration, Audiotactile Concert Perception.

1 Introduction

Measurements in real concert situations confirm the existence of whole body vibrations. If a bass drum is hit or the double bass plays a tone the perceived vibrations are noticeable. Nevertheless, in most cases the concert visitor will not recognize the vibrations as a separate event. The vibrotactile percept is integrated with the other senses (e.g. vision and hearing) to one multi-modal event.

Experiencing a concert, the listener expects vibrations, even if he is not aware of it all the time. These expected vibrations are missing in a traditional multimedia reproduction setup. According to Jekosch [1] the perceived quality of an entity (e.g. a reproduction system) results from the judgment of the perceived characteristics of an entity in comparison to its desired/expected characteristics.

M.E. Altinsoy, U. Jekosch, and S. Brewster (Eds.): HAID 2009, LNCS 5763, pp. 119–127, 2009.
© Springer-Verlag Berlin Heidelberg 2009

If in the reproduction situation the vibratory component is missing, there might be a loss of perceived quality, naturalness or presence of the concert experience. To look at it from the other side: The perceived quality of a conventional reproduction system might be improved by adding whole body vibrations. This study focuses on whole body vibrations for a seated person, like they are perceived in a classical chamber concert.

Unfortunately there is no vibration channel in conventional multimedia productions. Therefore it would be advantageous if the vibration signal could be generated using the information stored in the existing audio channels. This might be reasonable, since the correlation between sound and vibration is naturally high in everyday situations. The questions in focus of this study are:

1. Is it possible to generate a vibrotactile signal using the existing audio channels of a conventional 5.1 surround recording?
2. Up to which frequency should the WBV be reproduced?
3. Is there an ideal amplitude for the reproduction of WBV?

Previous studies are primarily concerned with the perception of synchrony between acoustical and vibratorical stimuli ([2], [3], [4], [5],[6]). Walker et al. [7] investigates the tactile perception during reproduction of action oriented DVD movies. In Merchel et al. [8] a pilot experiment is described, which aims at investigating the influence of WBV on the overall quality of the reproduction of concert DVDs. This paper describes an extended experiment with 20 subjects. The focus is on different vibration parameters like amplitude and frequency range of the vibration.

2   Experiment

2.1   Stimuli

The stimuli should include instruments for which low frequency vibrations and sounds are expected. Another criteria was that the stimuli represent typical concert situations for both classical and modern music. To place the subject in a standard multimedia reproduction context, an accompanying picture from the DVD can be projected. The video sequence shows the stage, the conductor or the individual instrumentalists while playing. The participant in the experiment should have enough time to become familiar with the stimulus. Thus a stimulus length of 1.5 minutes was chosen. The following sequences were selected:

- Bach, Toccata in D minor (*church organ*)
- Verdi, Messa Da Requiem, Dies Irae (*kettledrum, contrabass*)
- Blue Man Group, The Complex, Sing Along (*electric bass, percussion, kick drum*)
- Dvořák, Slavonic Dance No. 2 in E minor, op. 72 (*contrabass*)

The sum of the three frontal channels and the LFE channel was used to generate the vibration signal (see Figure 1). Two low pass frequencies were implemented

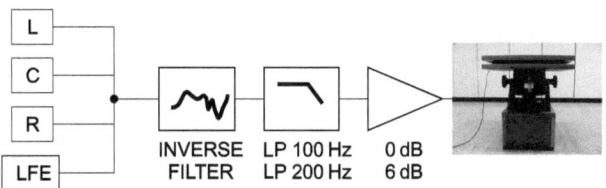

Fig. 1. Generation of the vibration signal using four out of six DVD audio channels. The transfer characteristics of the vibration chair was compensated and the signal was filtered with a variable low pass as well as variable amplification.

($f1 = 100\,\text{Hz}$, $f2 = 200\,\text{Hz}$), which seemed suitable for the chosen stimuli. The low pass filter was a steep Butterworth filter with 10th order. Two test persons adjusted the amplitude of the vibration, until it was just noticeable for all frequencies. This amplitude is further referred to as $a1 = 0\,\text{dB}$. To study the influence of the vibration amplitude, an additional signal with an $a2 = 6\,\text{dB}$ amplified vibration was generated. The peak value of the vibration at the surface of the seat in vertical direction was measured using an Endevco triaxial seat pad accelerometer. The peak vibration was measured between 0.25 and $0.60\,\text{m/s}^2$, depending on the reproduced sequence.

2.2 Setup

Figure 2 shows the used setup for reproduction of surround recordings according to ITU [9]. It was build in front of a silver screen for video projection. Five Genelec 8040A loudspeakers and a Genelec 7060B subwoofer were used. In addition vertical whole body vibrations have been reproduced using a self build electrodynamical vibration seat. The transfer characteristic of the shaker loaded with a seated person has been measured using an Endevco triaxial vibration pad. This frequency response depending on the individual test person is called the Body Related Transfer Function (BRTF) [10]. All stimuli have been compensated for the transfer characteristic of the seat in vertical direction by using inverse filters in MATLAB.

2.3 Crosstalk

There was also crosstalk between the different systems. The subwoofer excited some seat vibrations. They were measured to be below $0.5\,\text{mm/s}^2$. The peak vibrations reproduced with the shaker reached from 250 to $600\,\text{mm/s}^2$. This is factor 1000 above the sound induced vibrations. Thus it was concluded, that crosstalk is uncritical.

The vibration seat itself radiated some sound. It was measured at the hearing position to be 40 dB below the signal reproduced by the loudspeakers. The loudspeaker generated sound pressure level was approximately 68 dB(A).

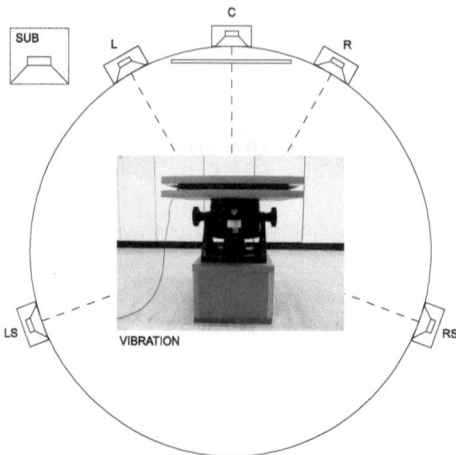

Fig. 2. Setup with six loudspeakers according to ITU [9]. An additional shaker was used to reproduce vertical whole body vibration.

2.4   Subjects

20 Subjects participated voluntarily in this experiment (15 male and 15 female). Most of them were students between 20 and 46 years old (mean 23 years) and between 48 and 115 kg (mean 75 kg). All stated to have no known hearing or spine damages. The average number of self reported concert visits per year was 18 and ranged from 1 to 100. The highest number was reported from one subject, who played guitar in a band. The preferred music styles were rock music (15 subjects), classic (9 subjects), pop (6 subjects), jazz (3 subjects) and 12 subjects preferred other genres in addition.

2.5   Experimental Design

Each subject had to judge 20 stimuli, five for each music sequence. All versions of one sequence were played on after the other, always starting with the no vibration condition. The remaining four combinations of low pass frequency ($f1 = 100\,\text{Hz}$, $f2 = 200\,\text{Hz}$) and amplitude ($a1 = 0\,\text{dB}$, $a2 = 6\,\text{dB}$) had been randomized between subjects using a balanced latin square, a williams square. The final presentation order is illustrated in Table 1.

Before starting with the experiment the subjects had to do a training with two stimuli to get familiar with the task and the stimuli range. The used stimuli was the first 1.5 minutes from Bizet - Carmen (Prelude), which is a classical composition with kettledrum and contrabass. After or while listening to the concert reproduction, the subject had to judge the overall quality of the concert experience using a quasi continuous scale. Verbal anchor points from bad to excellent have been added similar to the method described in ITU-T P.800 [11]. Figure 3 shows the used questionnaire. In addition presence and naturalness had to be evaluated by means of a five point Rohrmann scale [12].

Table 1. Order of presentation of all factor combinations to the first four subjects. The first stimuli in one sequence block is always without vibration. The other four combinations are randomized using a williams square. The presentation order of the music sequences are also randomized between subjects using a williams square.

Sequence	Bach					Verdi					BMG					Dvorak				
Vibration Amplitude	no	a1		a2		no	a1		a2		no	a1		a2		no	a1		a2	
Vibration Lowpass Freq.	no	f1	f2	f1	f2	no	f1	f2	f1	f2	no	f1	f2	f1	f2	no	f1	f2	f1	f2
Subject 1	1	2	3	5	4	6	8	9	7	10	16	19	20	18	17	11	15	12	14	13
Subject 2	11	13	14	12	15	1	4	5	3	2	6	10	7	9	8	16	17	18	20	19
Subject 3	16	19	20	18	17	11	15	12	14	13	1	2	3	5	4	6	8	9	7	10
Subject 4	6	10	7	9	8	16	17	18	20	19	11	13	14	12	15	1	4	5	3	2

...

Overall Quality

```
      Excellent     Good       Fair       Poor       Bad
    —|—+—+—+—|—+—+—+—|—+—+—+—|—+—+—+—|—
```

Fig. 3. Questionnaire to evaluate the overall quality of the concert experience

3 Results and Discussion

For statistical analysis the evaluation values were interpreted as numbers on a linear scale from 1 to 5. Data was checked for normal distribution with the KS-test. A multifactorial analysis of variance was carried out. The averaged results for the overall quality evaluation are plotted in Figure 4 with mean and 95% confidence intervals for all 20 stimuli. It can be seen that the influence of the different vibration parameters (low pass frequency $f1 = 100\,\text{Hz}$, $f2 = 200\,\text{Hz}$ and amplitude $a1 = 0\,\text{dB}$, $a2 = 6\,\text{dB}$) on the overall quality judgement is relative small.

Although reproduction with vibration is judged better than reproduction without vibration in most cases. This is illustrated in Figure 5. The quality evaluations for all vibration versions have been averaged for each music sequence. A t-test for paired samples showed very significant differences on a 1% significance level for all music sequences, except Dvořák. This music composition is very calm with gentle contrabass. Whole body vibrations might not be expected for this kind of stimuli. For all other sequences the vibration reproduction improved the perceived quality of the concert experience significantly. There was no significant influence of preferred music style on the evaluation of the reproduction.

There was no significant overall preference for a specific vibration amplitude or low pass frequency.

The interaction between the factors *low pass frequency* and *sequence* was significant on a 5% significance level. An interaction diagramm is plotted in Figure 6. It can be seen that for the sequences from Bach, Verdi and Dvořák it is preffered to reduce the frequency range for whole body vibrations to 100 Hz. If frequencies

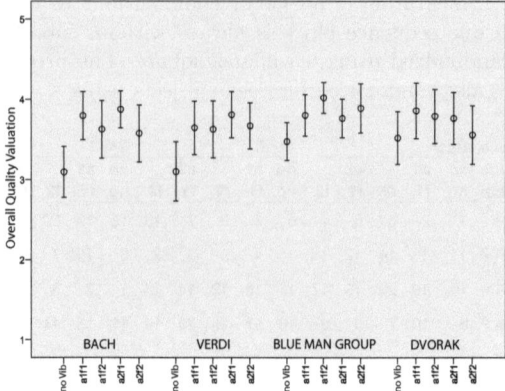

Fig. 4. Mean overall quality evaluation of all 20 participants with 95% confidence intervals

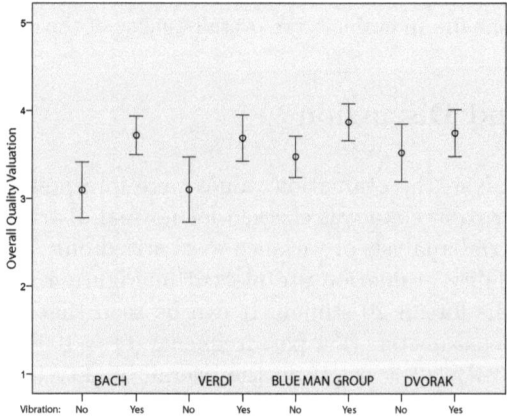

Fig. 5. Comparison of overall quality evaluation for reproduction with and without accompanying vibration plotted with 95% confidence intervals. It can be seen that reproduction with vibration ist judged better.

up to 200 Hz are reproduced the judgement is slightly worse. This might be due to a prickling sensation that is induced through higher frequency content. In addition no strong whole body vibrations might be expected in the range from 100 to 200 Hz for organ, kettledrum or contrabass. Contrary for the Blue Man Group sequence a 200 Hz low pass is favored.

Figure 7 helps to understand this result. The figure shows spectrograms which plot the frequency content (mono sum of L, C, R and LFE channel) over time for all four music sequences. For the church organ in Bachs Toccata in D minor (top left) the keynote and overtones are apparent. Note that the keynote remains below 100 Hz. In the Dvořák (bottom left) and Verdi (bottom right) sequence

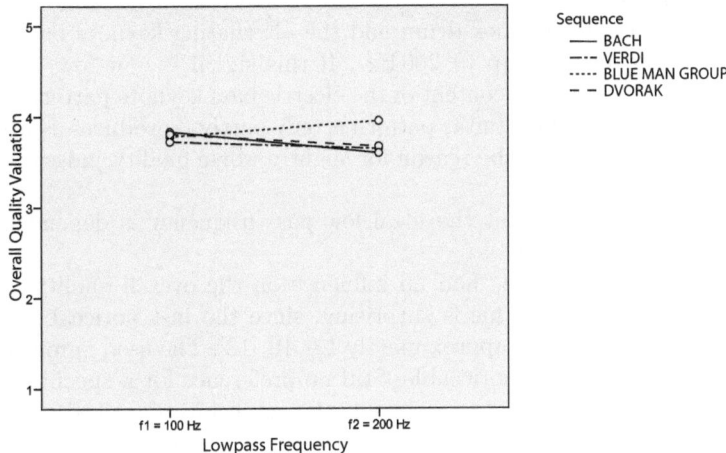

Fig. 6. Diagram showing the significant interaction between low pass frequency and sequence. The lower low pass frequency ($f1 = 100\,\text{Hz}$) is judged better in all cases except for the sequence Blue Man Group (BMG).

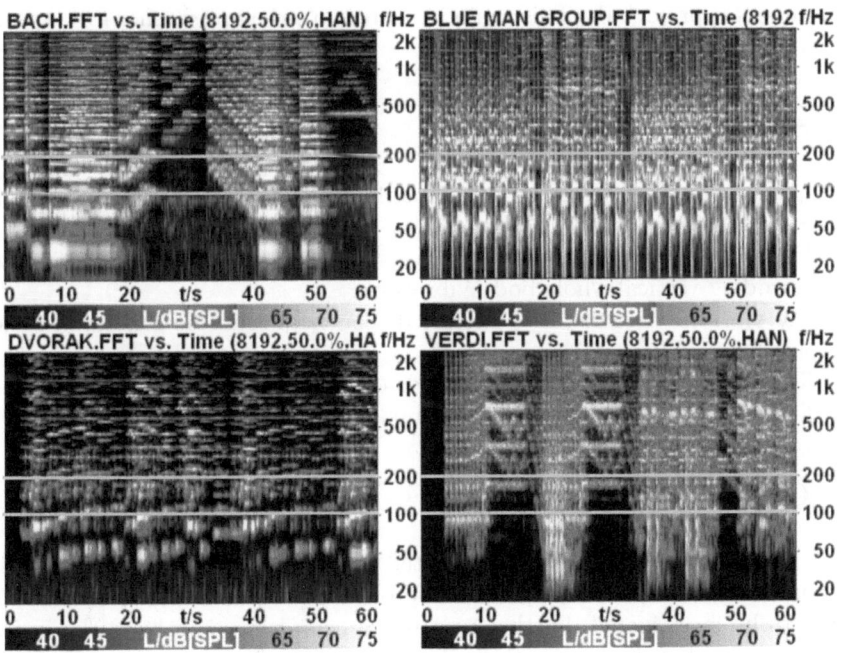

Fig. 7. Spectrograms of the mono sums for all four music sequences. The low pass frequencies (100 Hz and 200 Hz) are plotted with solid lines.

the keynotes remain as well below 100 Hz. In contrast in the Blue Man Group sequence (top right) the kick drum and the alternating keynote pattern of the bass guitar can be seen up to 200 Hz . If this signal is now low pass filtered with 100 Hz the frequency content of the electric bass keynote pattern is clipped. Thus the alternating bass guitar pattern is only partly reproduces as whole body vibration. This might be the reason for slightly worse quality judgments of this stimuli.

It can be concluded that the ideal low pass frequency is dependent on the particular music sequence.

The vibration amplitude had no influence on the overall quality judgments of the music sequences. This is surprising, since the just noticeable difference for vibration amplitude is approximately 1.5 dB [13]. The used amplitude difference of 6 dB was clearly noticeable. Still no preference for a specific amplitude was found. In an additional experiment, the subjects were asked to adjust the vibration amplitude for the same music sequences to an optimal level. A markerless infinite rotary knob (PowerMate, Griffin Technology) was used to avoid any visual cues. The results varied between subjects with a standard deviation of approx. 6 dB. This indicates a broad range of preferred vibration amplitudes for musically induced whole body vibrations. If one subject had to adjust the vibration amplitude for one stimuli repeatedly, the within subject standard deviation was approx. 4 dB. This again indicates no strong preference for a specific amplitude.

4   Summary

This paper investigates a reproduction method for whole body vibrations, which are generated from audio recordings.
  - The perceived overall quality of concert DVD reproduction can be improved by adding vertical whole body vibrations.
  - The vibration signals for the selected sequences could be generated by low pass filtering the audio sum signal.
  - The ideal low pass frequency depends on the specific music content.
  - The preference for a specific whole body vibration amplitude varies.

5   Outlook

This study uses a broad band calibrated whole body vibration reproduction system. The influence of an applicable reproduction solution in real life situations needs to be investigated. Further research is necessary to evaluate the general perception of whole body vibrations. The generation of the vibration signal using an audio recording is promising. However more complex processing than low pass filtering seems necessary. There are different approaches which are investigated at the moment.

References

1. Jekosch: Voice and speech quality perception - Assessment and Evaluation. Springer, Berlin (2005)
2. Altinsoy, Blauert, Treier: Intermodal effects of non-simultaneous stimulus presentation. In: Proceedings of the 7th International Congress on Acoustics, Rome, Italy (2001)
3. Daub, Altinsoy: Audiotactile simultaneity of musical-produced whole-body vibrations. In: Proc. of the Joint Congress CFA/DAGA, Strasbourg, France (2004)
4. Martens, Woszczyk: Perceived Synchrony in a Bimodal Display: Optimal Delay for Coordinated Auditory and Haptic Reproduction. In: ICAD, Sydney, Australia (2004)
5. Walker, Martens, Kim: Perception of Simultaneity and Detection of Asynchrony Between Audio and Structural Vibration in Multimodal Music Reproduction. In: AES 120th Convention, Paris, France (2006)
6. Altinsoy: Auditory-Tactile Interaction in Virtual Environments. Shaker Verlag, Aachen (2006)
7. Walker, K., Martens, W.L.: Perception of audio-generated and custom motion programs in multimedia display of action-oriented DVD films. In: McGookin, D., Brewster, S. (eds.) HAID 2006. LNCS, vol. 4129, pp. 1–11. Springer, Heidelberg (2006)
8. Merchel, Altinsoy: 5.1 oder 5.2 Surround - Ist Surround taktil erweiterbar? In: DAGA, Dresden, Germany (2008)
9. ITU-R BS.775-1: Multichannel stereophonic sound system with and without accompanying picture (1992)
10. Altinsoy, Merchel: BRTF - Body Related Transfer Functions for Whole-Body Vibration Reproduction Systems. In: DAGA, Rotterdam, Netherlands (2009)
11. ITU-T P.800: Methods for objective and subjective assessment of quality (1996)
12. Rohrmann: Empirische Studien zur Entwicklung von Antwortskalen für die sozialwissenschaftliche Forschung. Zeitschrift für Sozialpsychologie 9, 222–245 (1978)
13. Bellman, Remmers, Mellert: Basic Experiments on the Perception of Vertical Whole-Body Vibrations. VDI-Tagung Humanschwingungen, Darmstadt (2004)

The Effect of Multimodal Feedback Presented
via a Touch Screen on the Performance of Older Adults

Ju-Hwan Lee[1], Ellen Poliakoff[2], and Charles Spence[1]

[1] Crossmodal Research Laboratory, Department of Experimental Psychology,
University of Oxford, South Parks Road, Oxford, OX1 3UD, UK
[2] School of Psychological Science, University of Manchester, Coupland Building,
Manchester, M13 9PL, UK
{juhwan.lee,charles.spence}@psy.ox.ac.uk,
ellen.poliakoff@manchester.ac.uk

Abstract. Many IT devices – such as mobile phones and PDAs – have recently started to incorporate easy-to-use touch screens. There is an associated need for more effective user interfaces for touch screen devices that have a small screen area. One attempt to make such devices more effective and/or easy to use has come through the introduction of multimodal feedback from two or more sensory modalities. Multimodal feedback might provide even larger benefits to older adults who are often unfamiliar with recent developments in electronic devices, and may be suffering from the age-related degeneration of both cognitive and motor processes. Therefore, the beneficial effects associated with the use of multimodal feedback might be expected to be larger for older adults in perceptually and/or cognitively demanding situations. In the present study, we examined the potential benefits associated with the provision of multimodal feedback via a touch screen on older adults' performance in a demanding dual-task situation. We compared unimodal (visual) feedback with various combinations of multimodal (bimodal and trimodal) feedback. We also investigated the subjective difficulty of the task as a function of the type of feedback provided in order to evaluate qualitative usability issues. Overall, the results demonstrate that the presentation of multimodal feedback with auditory signals via a touch screen device results in enhanced performance and subjective benefits for older adults.

Keywords: Multimodal User Interface, Multimodal Feedback, Multimodal Interaction, Older Adults, Touch Screen.

1 Introduction

According to a news item in USA TODAY (June 21, 2007), touch screen phones are poised for rapid growth in the marketplace. In addition to Apple's iPhone®, many international electronics companies have recently launched touch screen phones as new cutting-edge user interfaces. The shipment of touch screens is projected to jump from less than 200,000 units in 2006 to more than 21 million units by 2012, with the bulk of the components going to mobile phones. USA Today quoted a maker of touch

M.E. Altinsoy, U. Jekosch, and S. Brewster (Eds.): HAID 2009, LNCS 5763, pp. 128–135, 2009.
© Springer-Verlag Berlin Heidelberg 2009

sensors as saying that: *"This new user interface will be like a tsunami, hitting an entire spectrum of devices"*. Unlike input devices such as the computer mouse that require translation from one plane of movement to another, that require extra space, and can have substantial summative movement time between different parts of the screen, touch screen user interfaces have a one-to-one relationship between the control and display, and often no additional training is necessary for their efficient use. Accordingly, touch screens are now being used extensively in a variety of application domains owing to the intuitiveness and ease of direct manipulation in use. Touch screens will also be beneficial to various user groups, in particular, for older adults (i.e., for those aged 65 and older), who may not be familiar with the recent developments in electronic devices, including complicated functions and structures and also have age-related degeneration in memory, sensory perception, and other aspects of cognitive and motor processing [1]. What is more, as the number of people over the age of 60 is expected to reach 1 billion by 2020, representing 22 percent of the world's population (according to 2006 UN world population prospects), they are likely to become a major group amongst IT consumers.

Meanwhile, previous research has suggested that one of the most important factors determining the usability of touch screen interfaces is the size of menu buttons on the screen. It has been shown that touch screens only provide significant benefits to users when the size of the buttons is made sufficiently large [2, 3]. However, although some applications involving touch screen interfaces, such as information kiosk displays and ATMs, have sufficient space on the screen, others such as mobile phones and PDAs have only limited screen space. Considering that various devices with multiple functions in the IT industry are focused on the miniaturization of portable smart phones including mobile phones and PDAs, the limitation of the screen space represents one of the most important challenges in the field of mobile user interface design. Moreover, this spatial limitation of touch screen devices may constitute a greater constraint for older adults than for younger adults. Consequently, researchers have attempted to integrate information from different sensory modalities in order to overcome the spatial limit of visual information displays. For instance, Gaver (1989) proposed that auditory confirmation might provide a more obvious form of feedback for users than only visual feedback [4], while Akamatsu and Sato (1994) demonstrated that tactile or force feedback can be effectively linked to information provided via a visual display, to give users the advantage of faster response times (RTs) and a more extended effective target area [5].

What is more, certain types of multimodal feedback can enhance performance in direct manipulation tasks consisting of a series of 'drag-and drops' using a mouse, while lowering self-reported mental demand [6]. In this context, the present study was designed to provide empirical evidence regarding the benefits of multimodal (or multisensory) interfaces specifically for older adults. We investigated this issue using a touch screen device, asking whether multimodal feedback can help older adults overcome the constraints of screen space in a dual task situation. RTs, error rates, and subjective ratings of type of feedback were measured.

We conducted an experiment with older adults comparing unimodal (visual), bimodal (auditory + visual, tactile + visual), and trimodal feedback (auditory + tactile + visual) in response to button click events while participants dialed a series of numbers into a touch screen mobile phone. We investigated whether different combinations of modalities of touch screen feedback were differentially effective in facilitating participants' behavioral performance under both single and dual-task conditions. The participants were placed in a dual task situation; they had to perform a visual recognition task (holding a picture in memory) while they carried out the touch screen mobile phone task. In addition to collecting objective measures of performance, we also measured the subjective difficulty of the task as a function of the type of feedback provided. It has been argued that the subjective evaluation of task difficulty may be just as important in terms of measuring (and evaluating) usability as objective behavioral performance measures [7].

2   Experiment

2.1   Methods

2.1.1   Participants
Thirteen older adults (9 female; mean age 73 years, age range 69-75 years) with an average of 6 years mobile phone experience and normal or corrected-to-normal vision and hearing took part in the experiment. All of the participants were right-handed by self-report and all gave their informed consent prior to taking part in the study. All of the participants were naïve as to the purpose of the study which took approximately 50 minutes to complete. They were paid £10 each for their participation.

2.1.2   Apparatus and Stimuli
Each participant was seated in an acoustically-isolated booth. An 8.4 inch LCD touch monitor with vibrotactile feedback (Immersion® TouchSense® Touchscreen Demonstrator; 60 Hz refresh rate) was used to present the mobile phone task (see the right panel of Fig. 1). Auditory feedback was presented from a loudspeaker cone situated directly behind the touch screen. Auditory feedback was synthesized at 16-bit & 44.1

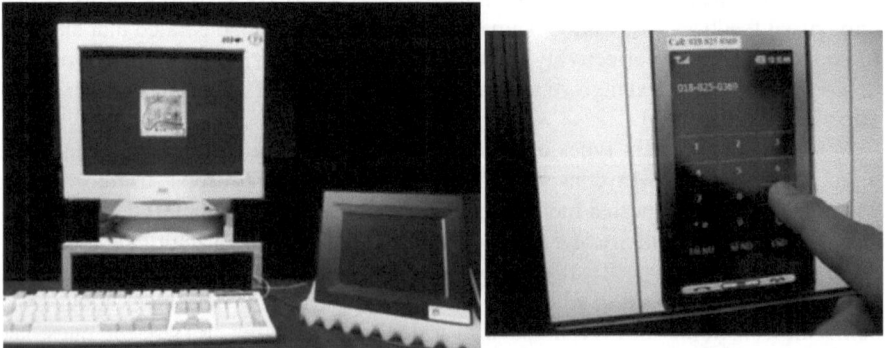

Fig. 1. The left panel shows the experimental set-up involving the secondary task of visual recognition and the primary task of using a mobile phone touch screen; the right panel shows an example of the touch screen device being used during the experimental mobile phone task

kHz and saved in WAV format. The auditory feedback consisted of a bell sound (main frequency: 355 Hz) presented for 150 ms at approximately 70 dB as measured from the participant's ear position. Tactile feedback (consisting of 50 Hz vibration) was presented via the touch screen for 50 ms from four actuators mounted within the touch screen device (index number 14 of the built-in Tactile Effects; for details, see Immersion® TouchSense® SDK Programming Guide, 2006). The tactile feedback was presented at a clearly supra-threshold level. In the dual task condition, the stimuli for the visual recognition task were presented on a 19 inch CRT monitor (75Hz refresh rate) placed 50 cm from the participant (see the left panel of Fig. 1). The picture stimuli were modern art pictures that could not easily be verbally encoded.

2.1.3 Experimental Design

There were two within-participants factors: Type of feedback: Visual only (V), tactile + visual (TV), auditory + visual (AV), or auditory + tactile + visual (ATV); Task: Single (mobile phone task) or dual task (mobile phone task + visual recognition task). For the single and dual task conditions, each participant completed 10 mobile phone task trials (consisting of five four-digit and five eleven-digit telephone numbers) with each type of feedback, with the order of presentation of the four types of feedback counterbalanced across participants. Each participant completed 80 mobile phone trials.

2.1.4 Procedure

In the single task condition, the participants only had to perform the mobile phone task; That is, they had to call a number on the touch screen device. On each trial, the telephone number to be typed was displayed on the touch screen itself. The participants were instructed to respond as rapidly and accurately as possible. If the participants made a mistake on the telephone task, they could delete their last key-press by pressing the 'clear' button on the keypad. In the dual task condition, the participants had to memorize a target picture that was presented in the middle of screen for two seconds. They were then required to hold the picture in memory while they executed the mobile phone task. Next, they had to pick out the target picture from amongst four alternatives. The participants were informed that they should divide their attention so as to perform both tasks as rapidly and accurately as possible. In addition, after the participants had completed 10 single and 10 dual task trials with each type of feedback, they completed a rating concerning the subjective difficulty of the immediately preceding feedback condition.

2.2 Results

Three performance measures were calculated: The mean accuracy of picture recognition, the mean accuracy and RT to complete the touch screen mobile phone task. In the visual recognition task, the number of correct responses did not differ significantly as a function of the type of feedback provided [$F(3,36)<1$, n.s.]. Participants were able to perform the visual recognition task with an average 77.3% correct (range from 75-79%) regardless of the type of feedback presented.

The RT and accuracy data from the touch screen mobile phone task (see Table 1) were combined into a single performance measure – inverse efficiency (IE) – that

Table 1. Mean RT and accuracy (S.D. in brackets) in the single and dual task conditions as a function of the feedback conditions

Task	Reaction Time (sec)		Accuracy (%)	
	Single	Dual	Single	Dual
V	8.9 (3.5)	9.5 (3.5)	83 (10)	86 (13)
TV	8.7 (2.4)	9.3 (2.9)	79 (17)	89 (19)
AV	7.8 (1.6)	8.2 (1.8)	93 (14)	99 (3)
ATV	7.6 (1.6)	7.7 (1.5)	90 (11)	95 (9)

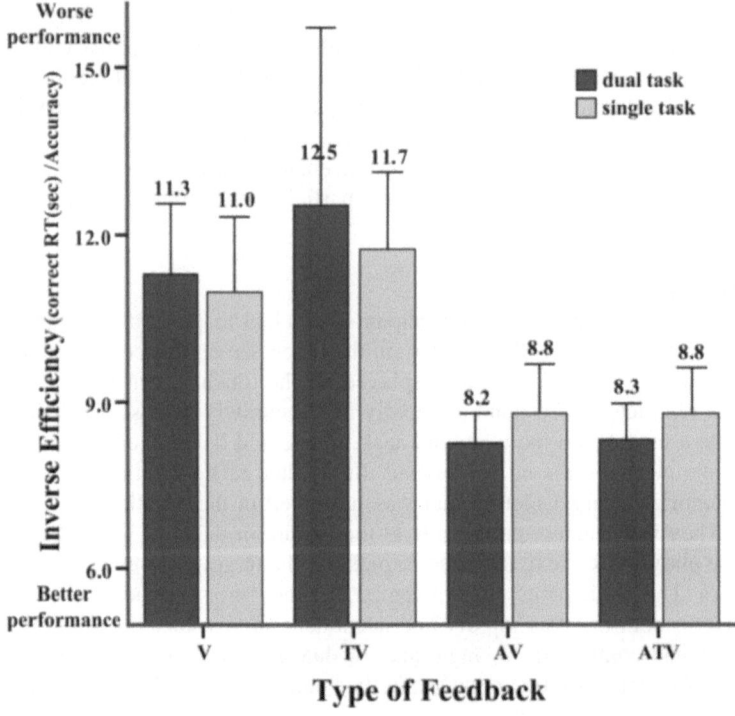

Fig. 2. Inverse efficiency (IE) scores on the mobile phone touch screen task plotted as a function of the type of feedback presented. The error bars show the standard errors of the means.

compensates for any potential speed-accuracy trade-off that may be present in one´s data [8]. IE = RT divided by the proportion of correct responses for a given condition on a participant; with a higher IE score indicating worse performance (see Fig. 2). Analysis of the IE data from the mobile phone task highlighted that the type of feedback affected performance significantly [$F(3,36)=3.10$, $p=.018$]. In particular, bimodal audiovisual and/or trimodal audio-visual-tactile feedback led to more efficient mobile phone performance than either unimodal visual and/or bimodal visuotactile

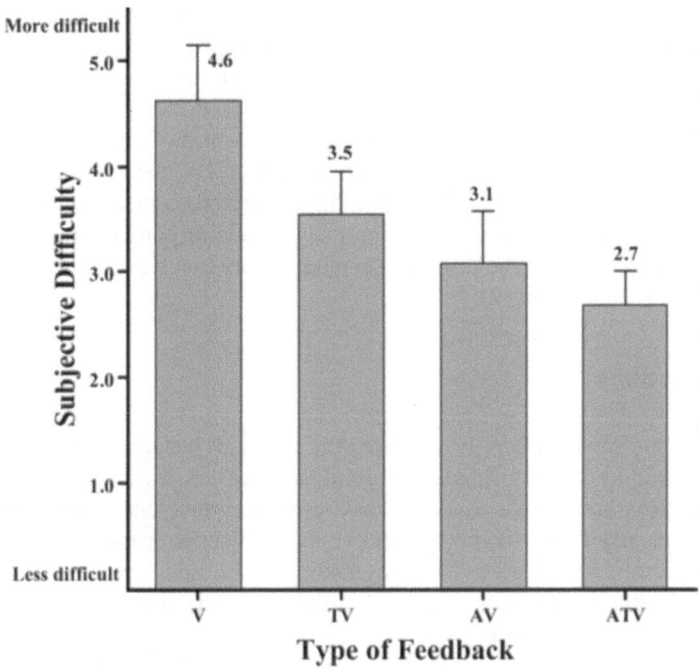

Fig. 3. Average Subjective Difficulty score (ranging from 1 ~ 5) plotted as a function of the type of feedback

feedback (see Fig. 2) [1-tailed Bonferroni comparisons (AV-V, ATV-V): p=.037, p=.045, respectively]. That is to say, crossmodal feedback involving auditory stimuli resulted in the older adults responding more efficiently. These results suggest that crossmodal auditory stimulation has a pronounced effect on participant´s performance of a touch screen task. There was, however, no significant main effect of task type (i.e., single vs. dual task) nor any interaction between the type of feedback and task. The reason that performance in the dual task condition was no worse than in the single condition may be due to the simplicity of the additional secondary visual recognition task that was used (remember that the response was collected after the completion of the mobile phone task). Nevertheless, the participants had to pay more attention (and devote more of their cognitive resources) in the dual task than in the single task trials. The results of the analysis of the behavioral data therefore demonstrate that participants were able to perform the mobile phone task more efficiently when they were given bi- or trimodal sensory feedback including auditory stimulation than when they were provided only with unimodal visual feedback or with bimodal visuotactile feedback.

A similar analysis of the subjective difficulty data revealed that the subjective difficulty of the mobile phone task also varied as a function of the feedback that was provided [$F(3,36)=6.32$, $p<.001$]. The subjective difficulty associated with trimodal feedback was significantly lower than that associated with either the unimodal visual or bimodal visuotactile feedback [1-tailed Bonferroni comparisons, p=.013; p=.044, respectively], while audiovisual bimodal feedback was no different from the other types of feedback (see Fig. 3). These results demonstrate that multimodal feedback (i.e., feedback that includes the stimulation of two or more of an interface operator's senses) can have a beneficial effect on subjective measures of difficulty, as well as on the more objective measures of participants' behavioural performance.

In summary, these results demonstrate the effectiveness of multimodal feedback presented via a touch screen and the importance of auditory information as a form of crossmodal stimulation in the task that seemingly only involves the visual and tactile modalities, for older adults [9].

3   Conclusions

The experiment reported here investigated the potential beneficial effect of the presentation of multimodal (as opposed to unimodal) sensory feedback on older adults' performance of a mobile phone dialing task, using a commercial touch screen device. Our results clearly show that both objective and subjective measures of older users' performance were enhanced by the presentation of bi- and trimodal (as opposed to unimodal) feedback including auditory stimulation. The experiment reported here tests a larger range of combinations of feedback modality than have been tested in previous research and does so in the practical context of the use of a touch screen mobile phone under both single and dual task conditions. Meanwhile, the absence of a very pronounced effect of vibrotactile feedback when delivered via a touch screen might be due to the increased possibility of slow movement and action/movement error (involving age-related kinematic differences) in the older than in younger adults [10]. However, the subjective ratings of task difficulty suggest that the addition of the tactile feedback had an effect in the trimodal condition, even though it was not picked up in the performance data.

Future research should therefore further investigate age differences, comparing any differential effects of multimodal feedback and cognitive workload on relatively old versus young adults. This is a particularly important issue given the rapid growth of older users of technology. There is some evidence to suggest that older people may find task-irrelevant multisensory stimuli harder to ignore than younger people [11]. On the other hand, it has also been argued that older people may, in fact, benefit more from multisensory (as opposed to unisensory) stimulation than younger people [12]. The relevant feature here then may be how relevant the additional sensory stimuli are to the performance of the participant's task.

Acknowledgements

We would like to thank Danny Grant and Immersion® for providing the vibrotactile feedback touch screen and the members of the Age and Cognitive Research Centre volunteer Panel for their time.

References

1. Rogers, W.A.: Individual differences, aging, and human factors: An overview. In: Fisk, A.D., Rogers, W.A. (eds.) Handbook of Human Factors and the Older Adult, pp. 151–170. Academic Press, London (1997)
2. Martin, G.L.: Configuring a numeric keypad for a touch screen. Ergonomics 31, 945–953 (1988)
3. Colle, H.A., Hiszem, K.J.: Standing at a kiosk: Effects of key size and spacing on touch screen numeric keypad performance and user preference. Ergonomics 47(13), 1406–1423 (2004)
4. Gaver, W.: The SonicFinder: An interface that uses auditory icons. Human Computer Interaction 4(1), 67–94 (1989)
5. Akamatsu, M., Sato, S.: A multi-modal mouse with tactile and force feedback. Int. J. Hu.-Com. St. 40, 443–453 (1994)
6. Vitense, H.S., Jacko, J.A., Emery, V.K.: Multimodal feedback: an assessment of performance and mental workload. Ergonomics 46, 68–87 (2003)
7. Hart, S., Staveland, L.: Development of NASA-TLX (Task Load Index): Results of empirical and theoretical research. In: Hancock, P.A., Meshkati, N. (eds.) Human mental workload, pp. 239–250. North-Holland, Amsterdam (1988)
8. Townsend, J.T., Ashby, F.G.: Stochastic modelling of elementary psychological processes. Cambridge University Press, London (1983)
9. Jacko, J.A., Emery, V.K., Edwards, P.J., Ashok, M., Barnard, L., Kongnakorn, T., Moloney, K.P., Sainfort, F.: The effects of multimodal feedback on older adults' task performance given varying levels of computer experience. Behav. Inform. Technol. 23(4), 247–264 (2004)
10. Ketcham, C.J., Seidler, R.D., Van Gemmert, A.W., Stelmach, G.E.: Age-related kinemetic differences as influenced by task difficulty, target size, and movement amplitude. J. Gerontol.: Psych. Sci. 57B(1), 54–64 (2002)
11. Poliakoff, E., Ashworth, S., Lowe, C., Spence, C.: Vision and touch in ageing: Crossmodal selective attention and visuotactile spatial interactions. Neuropsychologia 44, 507–517 (2006)
12. Laurienti, P.J., Burdette, J.H., Maldjian, J.A., Wallace, M.T.: Enhanced multisensory integration in older adults. Neurobiol. Aging 27, 1155–1163 (2006)

Audiotactile Feedback Design for Touch Screens

M. Ercan Altinsoy and Sebastian Merchel

Chair of Communication Acoustics, TU Dresden, Helmholtzstr. 10,
01069 Dresden, Germany
ercan.altinsoy@tu-dresden.de

Abstract. The use of touch sensitive displays and touch surfaces is just emerging and they are more and more replacing physical buttons. If a physical button is pressed, audio and tactile feedback confirms the successful operation. The loss of audiotactile feedback in touch sensitive interfaces might create higher input error rates and user dissatisfaction. Therefore the design and evaluation of suitable signals is necessary. In literature different researchers discuss implementation and evaluation of audio and tactile feedback for mobile applications using small vibration actuators, e.g. [1,..., 12]. However in ticket machines or automated teller machines the size of the actuator is not a limiting factor. Thus arbitrary vibratory stimuli can be generated. In this study, the tactile feedback is generated using an electro-dynamic exciter which allows amplitudes comparable to physical buttons. Real buttons normally produce multimodal feedback. Therefore multimodal interaction is an important issue for the touch screens. In this study, psychophysical experiments were conducted to investigate the design and interaction issues of auditory and tactile stimuli for touch sensitive displays and the combined influence of auditory and tactile information (i.e. vibration) on the system quality.

Keywords: Touch screens, multimodal interaction, auditory, haptic, evaluation, error rate.

1 Introduction

Haptic feedback brings the sense of touch (tactile sense) and force-feedback to multimedia applications in addition to the mostly utilised modalities i.e., the auditory and visual one. In many new applications like virtual reality, flight simulators and medical surgery, haptic modeling and simulation of different physical objects plays a pronounced role. Besides of the reality-based physical simulation, development of effects for perception-based intuitive haptic feedbacks is a big challenge.

The use of touchscreens, touch panels, and touch surfaces is growing rapidly because of their software flexibility and space and cost savings. They are more and more replacing physical buttons in different technical devices like mobile phones, hi-fi and TV-sets, navigation systems, ticket machines, cash-dispensers, etc. The big disadvantages of such kind of systems is the missing haptic feedback which is required for the confirmation of the successful operation. Several studies have concentrated on the technical solutions and new hardware issues for the haptic feedback implementation

M.E. Altinsoy, U. Jekosch, and S. Brewster (Eds.): HAID 2009, LNCS 5763, pp. 136–144, 2009.
© Springer-Verlag Berlin Heidelberg 2009

in mobile applications using small vibration actuators [1, ..., 12]. Pager motors, piezo-electric actuators, multi function transducers and electrotactile stimulators are possible actuators for such kind of systems. Hovewer the big disadvantages of these systems is that the actuators can not reproduce high amplitudes which is common for classical push-buttons, because of the restricted actuator size and the operating space. The studies of Hogan, Brewster and Johnson or Chang and O'Sullivan indicate that touch feedback in touchscreens increases not only productivity and make products easier to use, but also make user experiences more satisfying. In the study of Hogan, Brewster and Johnson, four different kinds of tactile stimuli (clicks) were evaluated regarding to their suitability to the different visual information and their perceived quality [3]. They have shown that by choosing congruent sets of audio/tactile feedback to be added to touchscreen visual buttons, not only are users' preconceptions of how the button should feel and sound met but also the perceived quality of the buttons is improved.

Several studies investigated the usability and the quality issues of ticket vending machines or automated teller machines [13, 14]. Most of the standard machines uses touch screen technology. One of the important error sources and quality problems for these machines is the missing tactile feedback similar to mobile phones. In ticket machines or automated teller machines the size of the actuator is not a limiting factor. Therefore it is possible to use big size actuators and reproduce high amplitude tactile feedbacks. In this study, the tactile feedback is generated using an electro-dynamic exciter which allows reproducing movement amplitudes comparable to physical push-buttons. Other advantages of the electro-dynamic exciter are the large frequency range and linear behavior.

The aim of this study is the evaluation of different haptic and auditory feedback signal forms and characteristics regarding their suitability to the touch screen applications. If auditory and tactile modalities are combined, the resulting multimodal percept may be a weaker, stronger, or an altogether different percept, and of course it is also possible that one modality can be dominant over the overall assessment related to the physical/perceptual ability, the nature of the task, or personal preference [15]. Besides of unimodal evaluations, further aim of this study is to gain a better understanding of the interaction of auditory and tactile information. In order to achieve these aims, experiments with unimodal and multimodal stimulus presentations were conducted and, especially, the effects of the perceptual discrepancy between the auditory and the tactile sensory modalities on the multi-sensory judgments were investigated.

2 Experiments

The first experiment is conducted to investigate the usability of tactile feedback. Different signal forms and characteristics are evaluated. The aim of the second experiment is to investigate the usability of the auditory feedback for touch screen applications. The third experiment is conducted to investigate audiotactile interaction effects.

2.1 Tactile Feedback Design for Touch Screens

2.1.1 Subjects
Six subjects, three male and three female, aged between 20 and 28 years, participated in this experiment. The subjects were undergraduate students and voluntarily participated in this study. All subjects were right handed and they used their right hand for the experiment. All subjects had self-reported normal hearing.

2.1.2 Experimental Set-Up
In this paper a touch sensitive system is presented that reproduces event triggered audiotactile feedback. The tactile component is generated using an electro-dynamic exciter, which is mounted behind a touch screen. The surface of the panel is divided into 6 virtual buttons. The experimental setup and the layout are shown in Figure 1.

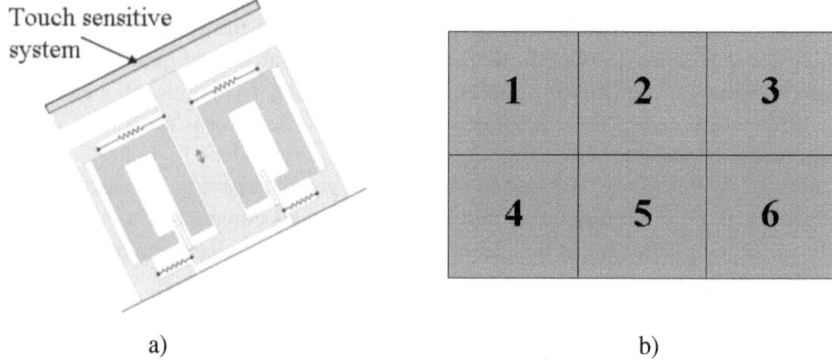

a)   b)

Fig. 1. a) The experimental setup: Touch sensitive system with electro-dynamic exciter b) Interface printed on the touch screen, it is divided into 6 virtual buttons (2 rows, 3 virtual buttons for each row)

2.1.3 Stimuli and Procedure
The goal of this experiment is to evaluate different tactile feedback forms regarding their suitability for touch screens, while pressing a virtual button on a panel. Five different stimuli, which can be seen in Figure 2 are selected (duration = 0.05 s each): sin, triangle, square, sin^2 and 50 Hz sinusoidal signal. The stimuli amplitude corresponds to the perpendicular displacement of the surface. Positive amplitude means movement towards the subject. The signal forms; sin, triangle, square and sin^2 show big similarity with the physical push-button feedback. Therefore 50 Hz sinusoidal signal, which doesn't show similarity with the physical push-button feedback, is additionally tested. The maximum amplitude of the stimuli is 4 mm. The tactile stimulus sin 50 Hz has an amplitude of 0.5 mm.

For the evaluation experiment, a dialing-numbers task is used. The participants are asked to dial 16 numbers displayed on an extra screen as fast and accurately as they can. The execution time and the errors during the task were measured. For each participant, the task is repeated 6 times. During each task the tactile feedback

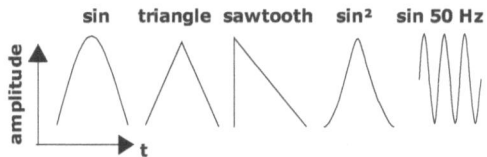

Fig. 2. Five different stimuli evaluated in this study. Length each = 0.05 s.

(5 stimuli used in the first experiment and 1 without any feedback) is the same for all 6 virtual buttons, but varies between different tasks.

After each task the participants were asked to evaluate the overall quality of the feedback, the suitability for confirmation and the comfort on a quasi continuous scale from "-5 (bad)" to "5 (good)". It was also possible to write down comments. The order of the stimuli was balanced between different participants.

2.1.4 Results

The performance of the subjects in terms of completion time and error rate of the dialing-numbers task is shown in Table 1. The difference between different feedback stimuli and none-feedback is significant for the number of errors and not significant in completion time. The number of errors of none-feedback is significantly larger than that of the other 5 kinds of feedback (Figure 3). The number of errors of sin feedback is significantly larger than that of sawtooth feedback.

Table 1. Performance for different tactile feedbacks showing mean values and standard deviations

	sin	triangle	sawtooth	sin ^2	sin 50 Hz	none	F	p	η
time to complete the dialing task in s	42.2±4.9	46.0±3.8	47.0±15.48	50.5±14.1	41.4±5.4	44.5±9.5	0.598	0.702	0.107
number of errors	1.0±1.1	1.3±1.5	0.2±0.4	1.3±1.7	0.9±0.7	4.5±2.5	8.030	0.000	0.616

To analyze the data of the evaluation part of the experiment (dialing-numbers task) ANOVA repeated measures were used. The results for the subjective valuation are shown in Table 2. Difference between different feedback stimuli and none-feedback is significant for overall quality and suitability for confirmation and almost significant for comfort of feedback.

Pairwise comparisons shows that the overall quality of none-feedback is significantly worse than that of the other five kinds of feedback, while they have no significant difference between themselves.

2.2 Auditory Feedback Design for Touch Screens

In this part of the study the touch sensitive system is used to reproduce event triggered audio feedback. Subjects and procedure were the same as in the first experiment.

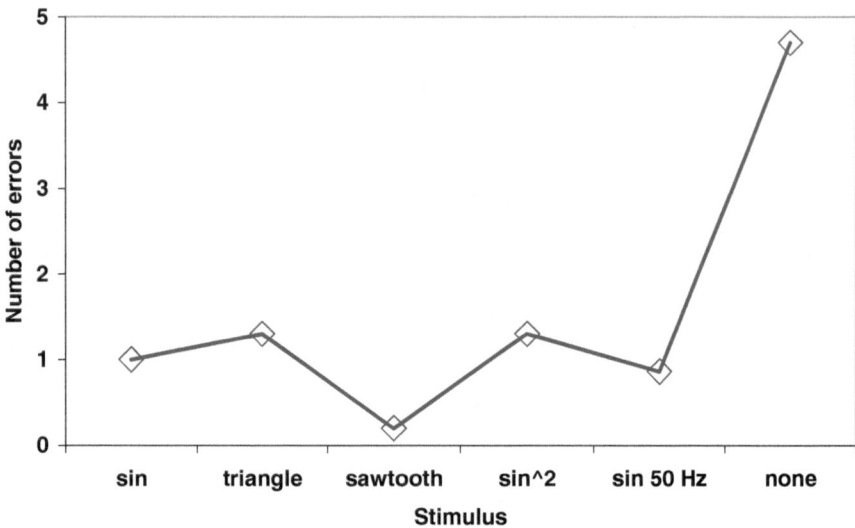

Fig. 3. Number of errors

Table 2. Perceptual quality ratings for different tactile feedbacks showing mean values and standard deviations

	sin	triangle	sawtooth	sin ^2	sin 50Hz	none	F	p	η
overall quality	2.5±1.52	2.2±1.47	2.7±2.25	2.3±1.6	2.5±0.9	-3.3±1.4	14.078	0.000	0.738
suitability for confirmation	2.5±1.64	2.5±1.64	2.8±1.60	2.2±1.8	2.5±1.4	-4.7±0.5	31.316	0.000	0.862
comfort of feedback	3.2±1.72	2.5±1.05	3.8±0.98	2.2±1.3	3.0±0.9	-1.3±3.6	4.382	0.051	0.467

Stimuli

The goal of this experiment is to evaluate different auditory feedback forms regarding their suitability for touch screens. Six different stimuli, which can be seen in Figure 4 are selected (duration = 0.05 s each).

Signal a, c and e are classical button sounds which have different frequency spectra and decay times. Particularly signal d has strong components at lower frequencies. Signal c has a broad band frequency spectrum with very short decay time. Signal a has some high frequency components. Signal b has a complex time sequence which is not common for classical push-button sounds. Signal d and f are DTMF (Dual-tone multi frequency) tones six and four which doesn't show any similarity with classical push-button sounds. The sound pressure level for all stimuli was 56 dB(A).

Results

The performance of the subjects in terms of completion time and error rate of the dialing-numbers task is shown in Table 3. The difference between different feedback stimuli is not significant for the number of errors and not significant in completion

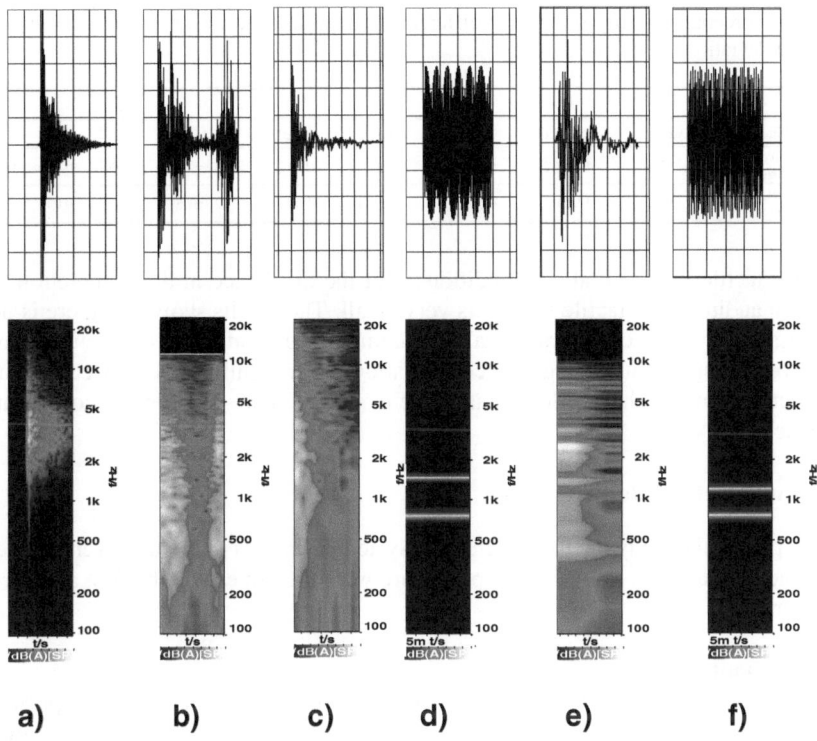

a)   b)   c)   d)   e)   f)

Fig. 4. Time sequences (above) and spectrograms (bottom) of six auditory stimuli. The color indicates the sound pressure level (yellow color: high sound pressure level, black color: low sound pressure level).

Table 3. Performance for different auditory stimuli (a,b,c,d,e and f) showing mean values and standard deviations

	a	b	c	d	e	f	F	p	η
time to complete the dialing task in s	45.0±3.4	44.6±3.9	47.4±2.6	47.4±3.9	48.9±3.9	44.2±4.3	6.598	0.651	0.607
number of errors	2.3±1.1	1.3±1.3	3.7±1.8	3.2±2.2	2.8±1.3	1.2±1.1	4.126	0.053	0.116

time. The number of errors for signals b and f are smaller than signals c,d,e and a. The reason can be that signals b and f are not typical button sounds. Particularly signal b has a characteristic time sequence.

To analyze the data of the evaluation part of the experiment (dialing-numbers task) ANOVA repeated measures were used. The results for the subjective valuation are shown in Table 4. The overall quality judgments seem very similar for different signal forms. Only signal c has a lower value than the others.

If tactile feedback only and auditory feedback only conditions are compared, it can be seen that The number of errors for auditory feedback alone conditions are larger

Table 4. Perceptual quality ratings for different auditory feedbacks showing mean values and standard deviations

	a	b	c	d	e	f	F	p	η
overall quality	2.9±0.7	2.8±1.1	1.5±2.6	2.6±1.3	2.8±0.9	3±1.3	16.103	0.000	0.766
suitability for confirmation	0.7±2.0	1.3±2.2	0.1±2.9	0.8±2.9	0.7±2.6	0.8±2.6	15.002	0.000	0.754
comfort of feedback	2.1±1.2	1.9±1.2	0.5±2.1	1.3±1.4	1.4±2.6	2.1±1.5	3.025	0.047	0.333

than for tactile feedback alone conditions. But the difference in the completion time between auditory and tactile signals is very small. The results show that there is not a big difference between overall quality judgments for auditory and tactile signals. However the suitability ratings for confirmation differ significantly between auditory and tactile signals. The tactile signals were evaluated more suitable for confirmation feedback.

2.3 Audiotactile Interaction

In this part of the study the touch sensitive system is used to reproduce event triggered audiotactile feedback. Subjects and procedure were the same as in the first and second experiment.

2.3.1 Stimuli
In this part of the study, two different tactile stimuli; $\sin^2$ and sin 50 Hz, are presented with different audio signals; signal a, b, c, d, e and f. The combined auditory with the tactile stimulus $\sin^2$ is called audiotactile 1 and the combined auditory with the tactile stimulus sin 50 Hz is called audiotactile 2.

2.3.2 Results
There is not a significant or important interaction effect regarding completion time and number of errors. The suitability for confirmation, the comfort ratings and the overall quality for audiotactile feedbacks are given in Table 5, 6, 7 and in Figure 5. The results of the audio only condition from the previous experiment are also given in the same tables for comparison.

The results show that if auditory signal is combined with the tactile signals, the tactile signal can alter the audio only ratings and almost all ratings increases. Particularly this effect is clearly observable for the ratings of the confirmation suitability

Table 5. *Confirmation suitability ratings* for different auditory and audiotactile feedbacks showing mean values and standard deviations (Reference: The rating of the tactile stimulus $\sin^2$ alone was 2.17±1.83 and the rating of the tactile stimulus sin 50 Hz alone was 2.5±1.5)

	a	b	c	d	e	f
only sound	0.7±1.9	1.3±2.2	0.1±2.9	0.8±2.9	0.7±2.6	0.8±2.6
audiotact. 1 (Tactile signal is **sin²**)	2.9±0.7	3.2±1.2	2.2±1.9	2.7±1.2	3.2±1.9	1.7±1.3
audiotact. 2 (tactile signal is **sin 50 Hz**)	3.5±0.8	2.4±1.8	2.1±1.0	3.4±1.3	2.0±1.6	2.9±1.7

Table 6. *Comfort ratings* for different auditory and audiotactile feedbacks showing mean values and standard deviations. (Reference: The rating of the tactile stimulus $\sin^2$ alone was 2.2±1.3 and the rating of the tactile stimulus sin 50 Hz alone was 3.0±0.9).

	a	b	c	d	e	f
only sound	2.0±1.2	1.9±1.2	0.5±2.1	1.3±1.4	1.4±2.6	2.1±1.5
audiotact. 1 (Tactile signal is $\sin^2$)	3.4±0.6	2.9±1.0	1.9±2.1	2.7±1.1	1.9±1.8	3.4±1.1
audiotact. 2 (tactile signal is **sin 50 Hz**)	2.8±0.6	2.6±0.7	1.9±2.6	2.8±0.7	1.7±1.3	2.5±1.5

Table 7. *Overall quality ratings* for different auditory and audiotactile feedbacks showing mean values and standard deviations. (Reference: The rating of the tactile stimulus $\sin^2$ alone was 2.3±1.6 and the rating of the tactile stimulus sin 50 Hz alone was 2.5±0.9).

	a	b	c	d	e	f
only sound	2.9±0.7	2.8±1.1	1.5±2.6	2.6±1.3	2.8±0.9	3±1.3
audiotact. 1 (Tactile signal is $\sin^2$)	3.2±1.9	3.1±2.2	2.0±2.9	2.8±2.9	3.0±2.6	3.5±2.6
audiotact. 2 (tactile signal is **sin 50 Hz**)	3.2±1.2	2.8±1.2	2.2±2.1	2.5±1.4	2.5±2.6	2.5±1.5

(Fig. 5). This tendency is also observed for comfort ratings, but also for overall quality ratings as a smaller effect. The tactile stimulus sin 50 Hz gives particularly very good confirmation suitability ratings, if it is combined with auditory signals a, d or f which have high frequency components.

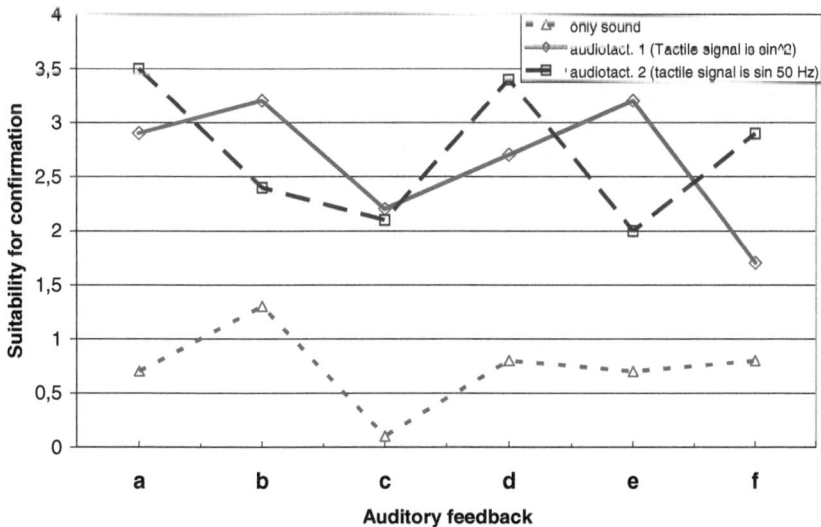

Fig. 5. Confirmation suitability ratings for audio only and audioctile feedback conditions

3 General Discussion and Conclusions

The results of this study show the advantage of tactile feedback in both perceptual quality and error rate for a number-dialing task compared to no feedback. The results indicate that completion time may not be a very good way to measure performance. Similar effects are also observed for the auditory feedback. Auditory and tactile feedbacks in touch screens make user experiences more satisfying than no feedback. The results of the audiotactile experiments show that if both modalities are combined, there are synergy effects. The tactile signal can improve the audio only ratings and almost all ratings get better.

In this study, the number of participants is small. Thus further experiments are necessary to investigate those hypotheses. Some further experiments are planned with higher amount of stimulus variety.

References

1. Chang, A., O'Sullivan, C.: Audio-Haptic Feedback in Mobile Phones. In: Proc. of the CHI 2005, Portland, USA (2005)
2. Tikka, V., Laitinen, P.: Designing haptic feedback for touch display: Experimental study of perceived intensity and integration of haptic and audio. In: McGookin, D., Brewster, S. (eds.) HAID 2006. LNCS, vol. 4129, pp. 36–44. Springer, Heidelberg (2006)
3. Hoggan, E., Brewster, S.A., Johnston, J.: Investigating the effectiveness of tactile feedback for mobile touchscreens. In: Proc. of the twenty-sixth annual SIGCHI conference on Human factors in computing systems, Florence, Italy (2008)
4. Doerrer, C., Werthschuetzky, R.: Simulating Push-Buttons Using a Haptic Display – Requirements on Force Resolution and Force-Displacement Curve. In: Proc. of EuroHaptics, Edinburgh (2002)
5. Shimoga, K.B.: A survey of perceptual feedback issues in dexterous telemanipulation. II. Finger touch feedback. In: Proc. of the IEEE Virtual Reality Annual International Symposium (1993)
6. Poupyrev, I., Maruyama, S., Rekimoto, J.: Ambient Touch: Designing Tactile Interfaces for Handheld Devices. In: Proc. of the 15th annual ACM symposium on User interface software and technology, Paris, France (2002)
7. Rosenberg, L.B., Riegel, J.R.: Haptic Feedback for Touchpads and other Touch Controls, United States Patent 7, 148, 875 B2 (2006)
8. Hayward, V., Alarcon, R., Rosenberg, L.B.: Haptic Pads for use with user-interface devices, United States Patent 7, 336, 266 B2 (2008)
9. Ling, S.-H.W., Chang, C.-C., Liao, W.-C., Lin, W.-C.: Method and Apparatus of Electrotactile Panel with Pointing System (2007)
10. Kaaresoja, T., Hemanus, J.: Mobile Phone Using Tactile Icons, United States Patent 6, 963, 762 B2 (2005)
11. Kim, S.H., Yang, Y.S., Lee, J.H.: Tactile and Visual Display Device, United States Patent Application, 2008/0129705 A1 (2008)
12. Nunokawa, K., et al.: Information Transmission Apparatus, Information Transmission Method, and Monitoring Apparatus, United States Patent, 7, 327, 277 B2 (2008)
13. Connell, I.W.: Error analysis of ticket vending machines: comparing analytic and empirical data. Ergonomics 41(7), 927–961 (1998)
14. Rogers, W.A., Gilbert, D.K., Cabrera, E.F.: An analysis of automatic teller machine usage by older adults: a structured interview approach. Applied Ergonomics 28, 173–180 (1997)
15. Altinsoy, E.: Auditory-Tactile Interaction in Virtual Environments. Shaker Verlag, Germany (2006)

Evaluation of User's Physical Experience in Full Body Interactive Games

Mitja Koštomaj[1,2] and Bojana Boh[2]

[1] Faculty of the Arts, Thames Valley University
Grove House, 1, The Grove, London W5 5DX, UK
[2] Faculty of Natural Sciences and Engineering, University of Ljubljana,
Aškerčeva 12, SI-1000 Ljubljana, Slovenia
mitja@kostomaj.net

Abstract. This paper is an evaluation of full body interactive games using Kroflič's and Laban's framework of Body, Space, Time and Relationship. An experiment with 8 participants playing 10 games for 20 minutes was conducted and recorded to digital video. Body, Space and Time elements have been measured using observation, motion tracking and Quantity of Motion (QoM). The results from the experiment informed the designer about the participants' physical experience through the analysis of postures used in each game, the quality of the movement, the body parts used in the interaction, the playing area, the direction of movement, direction of gaze, tempo, dynamics and QoM. The experiment informed the designer about important issues of the user's physical experience and proved that the method can provide useful information in the development and evaluation of full body interactive games. The theoretical work of Laban and Kroflič also proved useful for interaction and games design in the transition from desktop to full body interactive games.

1 Introduction

Although the start of full body interaction games goes back to the 1970s with Krueger's work in VideoPlace (Krueger, 2001), the design of full body interactive games is a new research and design area. Evaluation is complex, as the coordination of mental and physical skills is required for successful playing. Different methods, based mainly on the observation and written output have been used, while visualisations of human movement (ie. tracking) have been used in the analysis of sports and dance.

One of the first evaluation frameworks (Warren, 2003) was trying to answer how successfully the gamelets (i.e. sub elements of the games) incorporate the five important advantages of using computer vision over more traditional interfaces in video games which are Athleticism, Expressivity, Whole Body, Vocabulary of Action and Playability. Labanotation has been used in the interaction design context to evaluate Eyetoy Sony PlayStation (Sony Corporation, 2005), games (Loke, Larssen, Robertson, & Edwards, 2007), and it provided a valuable foundation for the design of movement-based interaction. Full-body interaction challenges for successful playing

M.E. Altinsoy, U. Jekosch, and S. Brewster (Eds.): HAID 2009, LNCS 5763, pp. 145–154, 2009.
© Springer-Verlag Berlin Heidelberg 2009

of Kick Ass Kung-fu (Höysniemi & Hämäläinen, 2005) requires (1) Locomotor skills, (2) Nonlocomotor skills, (3) Manipulative skills, (4) Movement awareness, (5) Body awareness, (6) Spatial awareness, (7) Focusing attention, (8) Observing visual feedback, and (9) Ability to remap movements. The above mentioned games have used different approaches, and their pioneering work is extremely valuable. It addresses issues of interaction and games design, such as the relation between the user movement and the games element on the screen. However, it does not analyse the user's physical experience.

Sports and health sciences and dance, on the other hand, have a great tradition of analysing human physical activity. Sports science analyses the performance of an individual player and a team. In professional sports this includes analysis of biomechanics, physiology, motor skills, game tactics, techniques and behaviour. Software such as Simi Motion (SIMI Reality Motion Systems GmbH, 2004) uses computer vision algorithms that perform motion analysis. Software Simi Scout (SIMI Reality Motion Systems GmbH, 2008) tracks the distance each player covers during a game. Speed and directions of movement, strokes and/or hits can also be measured. Lucent-Vision system (Pingali, Opalach, Jean, & Carlbom, 2001) tracks the motion trajectory of a tennis player. Data are visualized for further analysis with heat and coverage maps, speed charts, 3D images and animation, still and sequential imaging and more conventional graphs. Health experts are interested in what effect the full body and other physically interactive games have in respect of spiroergometry, heart rate and the whole body kinematics (Böhm, Hartmann, & Böhm, 2008).

In dance, Labanotation is a system of recording movement, and Laban Movement Analysis has been developed to analyse the movement. Originally devised by Rudolf Laban in the 1920's (Kroflič, 1992), it continues to be used in fields traditionally associated with the physical body, such as dance choreography, physical therapy and drama. Kroflič (Kroflič, 1992) based her own framework of Time, Body, Space and Relationship on Laban's work. The last category indicates that all categories are closely linked-up. In the EyesWeb (InfoMus Lab, 2008) project (Castellano, Bresin, Camurri, & Volpe, 2007), two expressive motion cues were considered: the Quantity of Motion (QoM), and the Contraction Index (CI). Both cues are global indicators of human movement; QoM is correlated with the user's energy, and CI with the space occupied by the user.

2 Materials and Methods

The Ambient Interactive Storybook (AIS) is an interactive storybook for an immersive environment, equipped with a web cam, a large visual display, and a PC using low level computer vision algorithms, such as detecting mass center and motion detection to measure the amount of motion. In AIS users spontaneously discover a story and characters, play games, solve problems, and express and reflect about their aesthetic experience with their body. The user experience is similar to Sony PlayStation's Eyetoy (Sony Corporation, 2005) games, but is targeting younger children. In ten full body interactive games (Koštomaj & Boh, 2009a), music sets the mood, but also influences the tempo of playing, as a higher music tempo usually requires from the user to move quicker and is therefore more physically demanding. Games are based on popular children's edutainment genres such as racing, matching, memory, catching and shooting, but are obviously adapted to the situation when the user sees him/her self on the screen.

Movement in AIS and in other full body interactive games differs from the movement in dance, physical therapy and drama, as it is based on open skills in an unpredictable environment. Even in a highly improvised dance one moves according to a plan (i.e. choreography), while in the AIS's full body interactive games it is the randomness of the game's elements that determines movement, which is therefore freer and less predictable. In this respect the movement is somehow similar to the movement in sports games. A good example is a basketball game, where there are many relationships between the players, space and the next action of a player, and the game is highly unpredictable.

Methods and participants - The experiment was conducted in Museum of Recent History, Celje, Slovenia in January 2008 with 8 participants aged between 5 and 13 years, and body height between 116 cm – 178 cm (Table 1:). Each participant played 10 games, 20 minutes each. The experiment was recorded with 2 video cameras.

3   Research Design

As no directly applicable research framework was found in previous works on sports and health sciences, Laban's and Kroflič's framework was used for the evaluation of the experiments. Before the start it was not clear what type of movement the observers could expect, therefore an open framework seemed to be the most appropriate.

Previous experiments using Laban's framework noted that it takes time to master Labanotation (Loke, Larssen, Robertson, & Edwards, 2007), but it "is a potentially useful tool to support the design of movement-based interaction". However, Labanotation was not used. Research was based on Laban's and Kroflič's framework with the research methods and strategies that are commonly used in games and interaction design (e.g. observations), as well as in sports science and dance (trajectory tracking, motion analysis, energy consumption, etc.).

In Laban Movement Analysis the **Body** category describes structural and physical characteristics of the human body while moving. This category is responsible for describing which body parts are moving, which parts are connected, which parts are influenced by others, and general statements about the body organisation. This area in sports science covers biomechanics with motion analysis and to some extent also the motor skills tests and physiology. According to the Laban Movement Analysis, the **Space** category identifies (1) kinesphere, which is the area that the body is moving within, and how the mover is paying attention to it; (2) Spatial Intention, the directions or points in space that the mover is identifying or using; (3) Geometrical observations of where the movement is being done, in terms of emphasis of directions, places in space, planar movement, etc. In the sports science research, one would conduct an analysis of the player position during the game, area covered, direction and speed of the movement, etc. The category **Time** covers Effort, or what Laban sometimes describes as dynamics, and what Labanotation calls "Length of time it takes to do the movement" (Kroflič, 1992). Sports and health science covers energy consumption through an endurance test using spiroergometry and monitoring the heart rate. The category **Relationship** indicates that all categories are closely linked-up and does not include any measurements. A good example is a measurement of energy consumption, which is strongly correlated with time and the area covered by a player.

3.1 Research Questions

The observations during the experiments focussed on the following research questions:

Body category: how many different postures a player uses during a game, what kind of a movement a player uses, how a player interacts with different parts of his/her body, how a player manipulates the body.

Space category: how much space a player needs during a game, in which directions a player moves, and what is the player's direction of gaze.

Time category: how a player responds to a certain dynamics and tempo of the game, and how much physical effort a player invests in the game.

3.2 Research Instruments

We found that using Laban's Kroflič's framework helps a designer to see a bigger picture of what needs to be analysed in the games. The research instruments, principles and methods were applied from both sports science, such as heat maps and coverage maps, and the dance, such as QoM and CI.

In the category **Body,** whole body kinematics or biomechanical measurements were not available for this experiment. A heat map was considered for the category body parts, but proved unreliable.

1. **The number of postures used in a game.** Each video was analysed, and the number of postures participants used in each game (i.e. standing, moving, lying, kneeling, running, crouching) was recorded.

2. **Quality of the movement.** Each video was analysed, and the observer gave a qualitative appraisal on the basis of how many different types of movement (i.e. standing, jumping, running, walking, bending, waving, crawling...) a participant used during the game.

3. **Body parts used in a game.** In the analysis of each video, the use of the participant's body parts was recorded in interaction with the game (i.e. trunk, legs, hips, arms and head).

In the category **Space,** a coverage map (Pingali, Opalach, Jean, & Carlbom, 2001) was chosen instead of the Contraction Index (Castellano, Bresin, Camurri, & Volpe, 2007), because it provided better and more accurate details. Direction of the movement answers to what Laban calls the spatial intentions and geometrical observations, and is a simplified version of the visualisation of routes (SIMI Reality Motion Systems GmbH, 2008). Gaze somehow complements the direction of movement, and was included in the analysis to find out whether a game required constant gaze toward the screen, or whether utilised sound and storytelling as a base for gameplay.

1. **Size of playing area.** Participant's left leg from the video (Image 1:) was tracked in each game to determine the size of the playing area.

2. **Direction of the movement.** Video and visualisations of tracking participant's left leg (Image 1:) were analysed to determine how many spatial lines (i.e. left and right, forward and backwards, circular, diagonal and standing) a participant used.

3. **Gaze.** Video was analysed to determine the ratio between the time a participant looked toward the screen, or when his/her gaze was turned away from it.

In the category **Time,** the tempo is defined as a speed of movement, and the dynamics as a rate of change in the speed of the movement. For example, the sprint is a high tempo game, and the long distance run a low tempo game. Although sprint has a high tempo, it also has a low level of dynamics, as the speed does not change greatly during the race. Basketball, on the other hand, is a game with a high level of dynamics, as the speed of playing tempo varies a lot. To measure a player's effort, equipment for spiroergometry (Böhm, Hartmann, & Böhm, 2008) was not available, therefore QoM (Camurri, Lagerlof, & Volpe, 2003) was used instead.

1. **Tempo.** Each video was analysed, and the observer gave a qualitative appraisal of the participant's tempo in each game.

2. **Dynamics.** Based on the video analysis, the observer gave a qualitative appraisal of the participant's dynamics in each game.

3. **QoM.** Each video was analysed using the Quantity of Motion (QoM) patch in the EyesWeb software (Camurri, Lagerlof, & Volpe, 2003). A mean for QoM for each participant's game was calculated.

4   Results with Discussion

4.1   General Observations from the Experiment

For successful playing, coordination of the child's physical and mental skills was required. All children were excited by the experience, and had put a lot of physical and mental effort into the 20-minute experience. They were deeply immersed into the story, into physical activities and the game environment, despite being watched and observed in a public space. Visual immersion was important, as it was exciting for the participants to see themselves on the screen while playing. Only the youngest participant (Participant H - Table 1) had a slow start and he needed more than one game to warm up, and fully understand and enjoy the experience. He had no previous experience of attending a sports or dance club, and only a limited experience of computers, unlike the majority of participants had. In developing a strategy for successful playing, participants used their previous knowledge and experiences of playing similar desktop computer games, and physical experience of exercising, sport, and/or dance. Children who were members of an athletics club would prefer to run; those who were in a basketball club would run forward and backward facing the screen all the time, while those from the dance club had used more expressive movement.

4.1.1   Body (Analysis of the Postures, Movement and Parts of the Body)

Analysis of the postures showed that participant's starting posture was standing. According to the tempo and the dynamics of the game, a participant used other postures, such as running, lying, kneeling, crouching, and others. Participants used more postures in games with a higher level of dynamics.

Analysis of the movement showed that in order to successfully interact with digital objects participants used movements such as: walking, running, jumping, bending, dancing, mimicking, drumming, rollerblading, hitting, crawling and standing still. More diverse movement was noted in games with a higher level of dynamics.

Using analysis by (Höysniemi & Hämäläinen, 2005) the following physical skills were identified: (1) locomotor skills (i.e. moving left and right to control avatars or objects, moving forward and backwards to reach the objects, moving forward to approach the camera or to readjust balance), (2) non-locomotor skills, such as speed (i.e. moving quickly to reach digital objects), balance (i.e. keeping balance after moving quickly), flexibility (i.e. bending to control an object), precision (i.e. patiently controlling objects), (3) manipulative skills (i.e. using limbs to control objects), (4) movement awareness (i.e. responding to the change of a digital object on the screen with a movement), (5) body awareness (i.e. knowledge of the position of the body and the digital objects), (6) focusing attention (i.e. distinguishing own image from digital objects), (7) observing visual and audio feedback (i.e. planing the next step).

Analysis of parts of the body used for interaction showed that most of the interaction was done with the body, less with arms, legs and head. Participants used more parts of the body in games with a higher level of dynamics. Specific interaction was noted when the story suggested a certain movement (e.g. to mimic a character or catch a star with the hand). A similar motion was noted by (D'Hooge & Goldsmith, 2001), who stated that 'When you see bubbles floating around you, your normal reaction is to pop the bubbles with your hands'.

In the category **Body,** not only the dynamics of the game influenced the movement, but also the rules of the game (e.g. to swim the river while avoiding all objects, or to catch certain objects), the interaction between the body and digital objects (e.g. to control the avatar left and right, slower or faster), the narrative component of the user interface (e.g. story and instructions that asked a participant to move in a certain way), the rhythm of the music, as a higher tempo required from participants to move faster, the visual feedback on the screen (i.e. a participant could see a projected image of himself/herself only within a limited distance on the left and right), and the physical space (i.e. size, lightning conditions and quality).

In this category the evaluation results helped the designer to understand how a particular full body interactive game, game genre or interaction offers an opportunity to induce a certain movement, such as smaller movements using genres that require patience, or expressive movements using storytelling.

4.1.2 Space (Analysis of the Playing Area, Directions of Movement and Gaze)

Analysis of the playing area was done by combining visualisations (Figure 1:) from different games by the same participant and from different participants of the same game. The results showed that games with a higher level of tempo and dynamics and a higher level of QoM required more space. Participants also used more space in games in which they did not look at the screen (i.e. gaze) for most of the playing time, as this allowed them to play out of the display boundaries.

Analysis of the direction of the movement showed that most of the movement was done in a left - right direction. Some intentional movement forward - backward was also noticed (i.e. in mimicking characters) and some unintentional movement forward - backward (i.e. when readjusting position after a participant lost balance in a game with a high level of dynamics). Some participants realized that if they moved closer to the camera, it was somehow easier to play. Participants used more directions in games with a higher tempo and in games where the rules did not require them to watch the screen most of time.

Analysis of the gaze pointed out that the participants observed the action on the screen most of the time. During the games with less strict rules and which did not require constant focusing on the screen, participants left the screen, used more space and moved in different directions, including in circles and diagonally.

Results of the **Space** category showed that the physical space is an important part of physical activity, as the size of playing area, objects (e.g. furniture), quality of floors (e.g. it is easier and safer to move on wooden floors) and lighting conditions in the space (e.g. light improves playing space) influenced the user's experience.

These results also informed the designer about the required size of the playing area. In Figure 2: all visualisations from game 9 were combined. An average distance for each participant in the game was established, as presented in Table 1:.

Table 1. An average distance from the screen in game 9 and participant's characteristics: sex (m for male, f for female), age, and body height

Participant	A f	B f	C f	D f	E m	F m	G m	H m
Age (years)	14	11	9	8	8	6	8	5
Body height (cm)	178	149	140	135	135	116	131	125
Distance (cm)	430	400	350	290	250	240	230	200

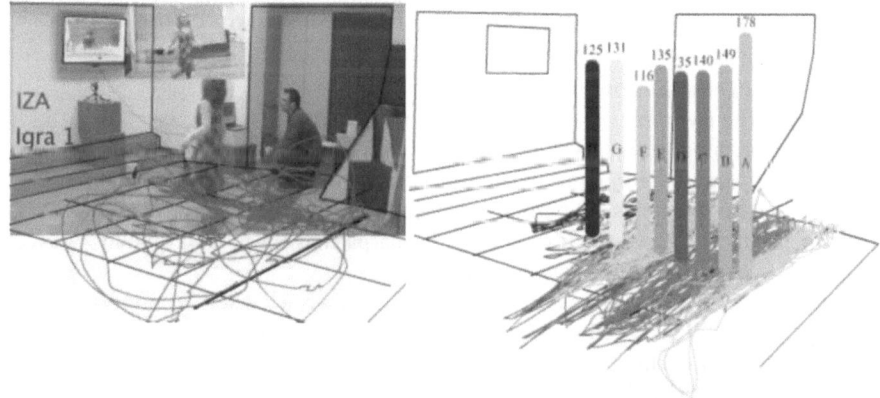

Fig. 1. Tracking participant's left leg and determining the playing area   **Fig. 2.** Comparing participants' movement and body height in game 9

Left - right was the dominant direction of the movement in the experiment. Such movement was also observed in most of the previous full body interactive games that used low level computer vision algorithms, such as Me2Cam (D'Hooge & Goldsmith, 2001), Sony Eyetoy (Loke, Larssen, Robertson, & Edwards, 2007) and QuiQui's Giant Bounce (Höysniemi, Hämäläinen, & Turkki, 2004). Left-right dominant movement was used by the authors in the design of educational full body interactive games, where players learned by making connections between the position of their body and the position of the digital objects in the direction left - right on the screen (Koštomaj & Boh, 2009b).

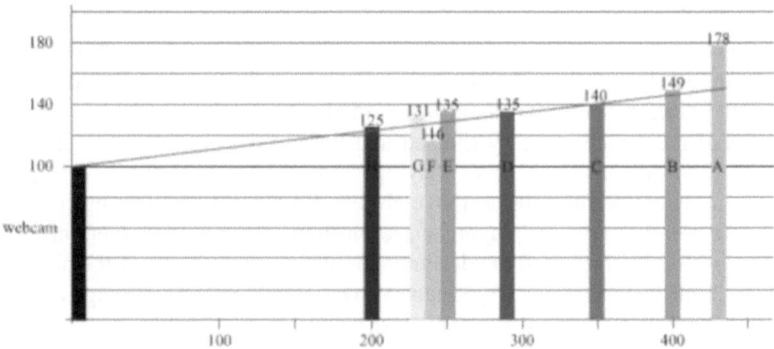

Fig. 3. Comparing participants' height and distance from the camera

As most of the computer games require constant focus on the screen, one of the interesting findings of this experiment was that games with less strict rules did not require the participants to watch the screen all the time. When they turned away, they could explore the playing area outside the screen, and therefore used more space, making the playing area bigger.

In the category Space, the main challenge for the designer is to create the environment, and to maximize the space as much as possible, to allow the physical activities that require more space (e.g. running, jumping).

4.1.3   Time (Tempo, Dynamics and QoM)

Analysis of the dynamics showed that a higher level of dynamics was noticed in the racing and catching genres, while lower levels of dynamics occurred in the matching and memory games and in games that required repetitive movements. In games with a higher level of dynamics, a participant used more postures, movement was more diverse, and more space was used. Sometimes a participant moved quickly, waited a couple of seconds, and then moved again.

Analysis of the tempo showed a higher tempo in games where a motion detection was used to define the speed of the game. A high level of tempo was also found in the catching game. In the high tempo games participants used more of the playing area.

Analysis of the QoM. A high value of QoM was found in games where the tempo of the game required a lot of movement with the whole body and where the participants used more space, moving in different directions. Low value of QoM was found in games where the participants did not observe the action on the screen all the time. In such games many had left the screen for some seconds and in that time the value of QoM was zero. Low value of QoM was also found in games that took a longer time to finish; in such cases participants put in less effort. A correlation between the energy consumption and QoM has been found (Camurri, Mazzarino, & Volpi, 2003). However, the results from the experiment indicated that QoM demonstrates a participant's physical effort, not his/her energy consumption. More work needs to be done in this area to establish a correlation between QoM and scientific measurements of energy consumption.

Many early commercial full body interactive games for Sony Eyetoy had a high tempo. Players had a high energy consumption, which resulted in exhaustion and a negative user experience. Most game genres trigger a competitive side in the player; if the player wants to achieve a better score, he/she needs to make a greater physical effort. An interesting approach was used in 'Ere Be Dragons (Active Ingredient, 2007), where a heart beat monitor was introduced, which allowed playing the game only when the player's rate was inside the healthy heart rates. In our experiment, slowing down was achieved by introducing a physical activity which required accuracy, smaller and finer motor skills; by manipulating motion detection algorithms to use their values for the speed of the elements in the game; by using music with lower tempo; by using genres such as memory or matching; and by making digital objects smaller or less visible.

The category **Time** informed the designer about how to use the length, tempo and dynamics of the game, with game elements such as music, gameplay, genres, to manipulate the participant's effort.

5 Conclusions

This research has been designed for better understanding on how to design a user's physical experience in full body interactive games. In the evaluation, elements important for the designer have been derived from Kroflič's and Laban's framework Body, Space, Time and Relationship, supported by sport science measurements. Findings from the experiment were not only useful in identifying the elements of the user's physical experience in full body interactive games, but also helped the designer understand correlations between the measurements.

A summary of results shows that:

(1) The diversity of the movement is defined by the dynamics and rules of the game, interaction between the body and digital objects, narrative component, rhythm of the music, visual feedback on the screen, and the physical space.

(2) Physical space is an important part of physical activity, as the size of playing area, objects, quality of floors and lighting conditions in the space influence the user's experience.

(3) The suggested width for the playing area is 4.50 m in depth (from the screen) and 5.50 m in width.

(4) The playing area can be bigger if games do not require a constant gaze directed to the screen. In such games audio and other haptic interfaces can play important role.

(5) The EyesWeb's QoM patch can be used for measuring participant's physical effort, but not energy consumption.

(6) Tempo and dynamics can manipulate user's physical effort by introducing movement that requires smaller and finer motor skills; by manipulating motion detection algorithms, by using music with lower tempo; by using suitable genres and by making digital objects less visible.

Studying Laban's work proved to be useful for informing interaction and games designers about issues important to the human body and human motion, and for making the transition towards the full body interactive games smoother. This knowledge has also opened the communication channels to experts in the fields of drama, dance and sports. The results will be used in further theoretical work to establish the components of user experience in full body interactive games and in further developments of educational and edutainment full body interactive games and stories.

Acknowledgement

Authors would like to thank Jožica Trateški, Tone Kregar and Andreja Rihter from The Museum of Recent History, Celje, Slovenia. Special thanks to Tara, Tea, Pia, Maša, Iza, Lan, Nik, Jaka and Žiga and their parents.

References

Active Ingredient: Ere Be Dragons (2007),
    http://www.i-am-ai.net/erebedragons/ (retrieved November 26, 2008)
Böhm, H., Hartmann, M., Böhm, B.: Predictors of metabolic energy expenditure from body acceleration and mechanical energies in new generation active computer games. In: Dagstuhl Seminar Proceedings, Munich, Germany (2008)
Camurri, A., Lagerlof, I., Volpe, G.: Emotions and cue extraction from dance movements. Intl. Journal of Human Computer Studies 59(1-2), 213–225 (2003)
Castellano, G., Bresin, R., Camurri, A., Volpe, G.: Expressive Control of Music and Visual Media by Full-Body Movement. In: Proc. of the 2007 Conference on New Interfaces for Musical Expression (NIME 2007), New York, pp. 390–391 (2007)
D'Hooge, H., Goldsmith, M.: Game Design Principles for the Intel® Play™ Me2Cam* Virtual Game System. Intel Technology Journal 4, 1–9 (2001)
Höysniemi, J., Hämäläinen, P.: Children's and parents' perception of full-body interaction and violence in a martial arts game. In: Proceedings of the 2005 conference on Designing for User eXperience, pp. 2–16. AIGA, San Francisco (2005)
InfoMus Lab. EyesWeb, Genova, Italy (2008)
Koštomaj, M., Boh, B.: Ambient Interactive Storybook, Unpublished report: University of Ljubljana (2009a),
    http://www.kostomaj.net/articles/KostomajBohAIS09.pdf
    (retrieved June 20, 2009)
Koštomaj, M., Boh, B.: Measuring impact on learning and motor skills in full body interactive games. Unpublished study. University of Ljubljana (2009b),
    http://www.kostomaj.net/articles/KostomajBohMotor09.pdf (retrieved June 19, 2009)
Kroflič, B.: Ustvarjanje skozi gib. Znanstveno in publicistično središče, Ljubljana (1992)
Krueger, M.W.: Responsive Environments - 1977. In: Paker, R., Jordan, K. (eds.) Multimedia: from Wagner to Virtual Reality. Norton, New York (2001)
Loke, L., Larssen, A.T., Robertson, T., Edwards, J.: Understanding movement for interaction design: frameworks and approaches. Pers. Ubiquit. Comput., 691–701 (2007)
Pingali, G., Opalach, A., Jean, Y., Carlbom, I.: Visualization of Sports using Motion Trajectories: Providing Insights into Performance, Style, and Strategy. In: 12th IEEE Visualization Conference (Vis 2001), pp. 75–82. IEEE, San Diego (2001)
SIMI Reality Motion Systems GmbH. Simi Motion, Unterschleissheim, Germany (2004)
SIMI Reality Motion Systems GmbH. Simi Scout, Unterschleissheim, Germany (2008)
Sony Corporation. Eyetoy.com (Sony), Sony PlayStation Eyetoy (2005),
    http://www.eyetoy.com (retrieved September 20, 2008)
Warren, J.: Unencumbered Full Body Interaction in Video Games (April 2003),
    http://a.parsons.edu/~jonah/jonah_thesis.pdf (retrieved October 1, 2008)

A Tangible Game as Distributable Platform for Psychophysical Examination

Matthias Rath[1] and Sascha Bienert[2]

[1] Technische Universität Berlin, Deutsche Telekom Laboratories,
Ernst-Reuter-Platz 7, 10587 Berlin
matthias.rath@tu-berlin.de
[2] Technische Universität Berlin, Institut für Sprache und Kommunikation, 10623 Berlin
sascha_bienert@gmx.de

Abstract. Through the use of built-in accelerometers a game-software for recent generation MacBooks allows control of a scenario of virtual moving objects by tilting the computer. Together with integrated visual and continuous auditory feedback from models based on physical principles the software forms a possible platform for online collection of psychophysical data.

1 Background and Motivation

The last decades have seen an increasing number of studies concerned with auditory perception of physical–ecological information (such as material [4][11] or size [1]). Most such works deal with discrete sets of information perceived in discrete auditory events, while the roll of *continuous* sonic feedback in as well continuous gestural human interaction with physical artefacts — despite its prominence in the "real" everyday world — has received less attention so far.

One example for the latter case is given by work conducted at an experimental tangible audio-visual interface, "*Ballancer*" [8], with central focus on continuous, seamless auditory perception of information of *velocity* of a virtual rolling ball. In short, in experiments at the *Ballancer* test subjects' performance in a target reaching task has improved under the presence of different types of continuous auditory feedback [9]. An important element of these studies and interface is a sound synthesis algorithm based on physical description of scenarios of contacting solid objects [6]. By using this sound model, apart from velocity information several other physical/ecological attributes may be expressed through sound feedback, such as the weight and diameter of the virtual ball, the roughness (or in general: structure) of the contacting surfaces, the resonance behavior of the supporting plane, which is mainly influenced by its material, and the hardness/elasticity of the rolling object (again a material property).

Tendentially such physical properties are most often and reliably perceived through the tactile channel (by touching the involved objects) but the sound emitted at contact such as rolling may also be an important cue (compare also e.g. [5]). In particular, the possiblity to support limited, or compensate for missing, tactile information through sound — in the line of thinking of *sensory substitution* — may be attractive for practical applications where possibilities of providing tactile feedback are restricted

M.E. Altinsoy, U. Jekosch, and S. Brewster (Eds.): HAID 2009, LNCS 5763, pp. 155–164, 2009.
© Springer-Verlag Berlin Heidelberg 2009

(e.g. for reasons of technology or display space). It is therefore one current motivation of the authors to exploit the metaphor and techniques behind the *Ballancer* — balanced rolling objects with continuous reactive auditory sound feedback — for examinations of auditory perception of object and surface properties.

Another central idea behind the implementation described in this contribution consists in gaining exeprimental data by means of a freely distributed game/demo-application software: the authors' works at the previous *Ballancer* interface showed that acquisition of human control movement data of sufficient amount to allow for statistically significant conclusions is a tedious task when relying on a laboratory environment. The wide distribution of sensor hardware (such accelerometers) in many devices today might here offer an attractive alternative. In the following we describe the main elements of a game/demo-application-like software which will form the basis for future perceptual studies. A first version of the application may be downloaded by now. It is the authors' intention to discuss the possible potential of such an easily distributable experimental game-like platform, share ideas and facilitate possibilties of collaboration.

2   Interface, Metaphor and Program Structure

As described, the intention behind the software presented in this contribution is to supply an easily distributable platform for psychophysical experiments, however appearing with a possibly "game-like" character. A general challenge was therefore to combine tangible access, a widespread underlying hardware platform, reactive continuous auditory feedback, precise synchronization of in- and output in different sensory channels, computational costs, and in particular the possibly conflicting aspects of experimental focus vs. fun of use.

For gestural user input with possibly widespread tangible devices a convenient decision is to rely on the use of accelerometers, as such sensors are today found in many portable devices such as mobile phones (e.g. Apple iPhone, Nokia N95), the Nintendo Wii game controllers and notebook computers. The presence of accelerometers in portable computers generally serves the aim to protect harddrives rather than purposes of gestural input so that no device-independant APIs for their use as input mode are available. For the last generation Apple *MacBooks* however, data from the built-in 3-d accelerometers may be accessed in a relatively convenient way, across the different models of these series (e.g. by using the open source "Unimotion" library). This was one reason for us to choose this platform for our first implementation, along with the facts of up-to-date computing power and graphical display facilities as well as a good quality sound output and relatively low-latency (as compared to much of other standard PC hardware) driver environment. Also, the implementation under Mac OS X should facilitate porting of the code to the iPhone platform, as far as computational power of this device will be sufficient for our realtime physics-based sound generation algorithms. (Work in this direction has recently been launched.)

2.1   Game Concept

As to our game, continuous gestural control interaction and sound feedback is the main focus. The accelerometer gives us the opportunity to determine the inclination of the

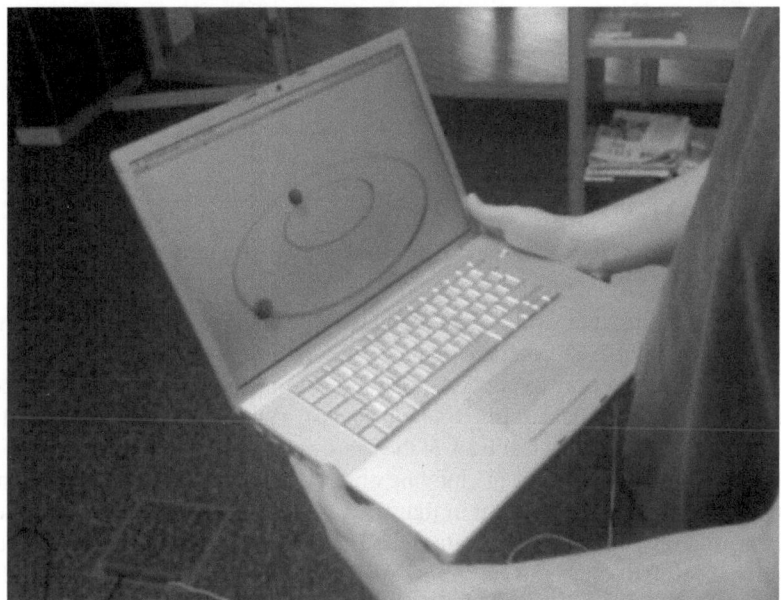

Fig. 1. The *orbits* game being played on a *MacBook*

MacBook. In this manner it can be used as an input device for steering tasks in human-machine interaction. One apparent scenario is to route an object on a plane or on a track. Indeed, there are quite a few games where the user has to balance an object (mostly a ball) in a maze-like virtual world[1].

As already mentioned we want to establish a platform that facilitates psychomechanical tests. In particular our application is geared towards sound feedback and gestural control. We thus chose the scenario of objects tied to circular tracks moving under the effects of gravity and friction. Tilting the tangible device causes a period of oscillation around the equilibrium, until due to friction the objects eventually come to rest. Each object can be steered to any place on the circle by performing adequate balancing movements. The player can also make the object rotate in a circular manner on the track.

Thus sound feedback is the most important feature of our application and explorations of the influence of diverse types of acoustic feedback on user experience and human performance in control tasks is one central interest. The sound feedback is computed based on physical considerations and put out as continuous realtime feedback of the objects movement. In the current configuration we provide a number of differing audio feedback basing upon two different algorithms: a *rolling sound model* and a *sliding sound model*. The sound feedback can be varied depending on parameters like size or mass of the object and resonance behaviour or surface profile of the underground.

In our first realization the scenario consists of two separate tracks with two independently moving objects. Both circular tracks contain target areas, but the challenge does

---

[1] For example common marble games like "Labyrinth" (iPhone), "Super Monkey Ball" (iPhone, Wii), "Kororinpa" (Wii).

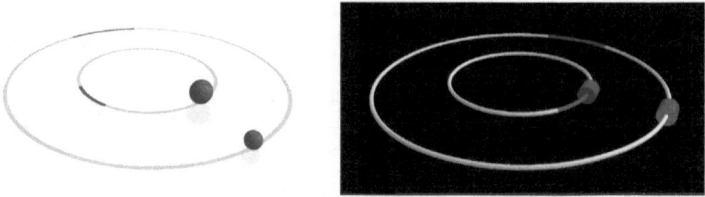

Fig. 2. Screenshots of two visual elements of the *orbits* game

not consist in finding them since they are clearly visible and finding a path to reach these areas is trivial; please compare figure 2.

Using two tracks not only supplies possibilities to create different tasks to accomplish for the user (as will be described below) but also enables us to present two different types of continuous auditory feedback at the same time.

For the visual appearance of the moving objects in our game we chose as first elements the very simple cases shown in figure 2: a ball which is rolling on a circle and a cylinder which is sliding on a circular rail. The object that we mainly use is the ball (primarily because of our focus on rolling sound). Some other sounds (e.g. the sliding sound) may seem unsuitable for the sonification of a ball's movement, a supposition that can be explored by means of the application. In this regard the sliding cylinders can give an impression of the interdependency of auditive and visual perception (for example by altering the objects but not the sound model or vice versa). Besides their function to visually maintain (or contradict) auditory feedback the cylinders are also useful to underline a stronger degree of friction.

As game logic and complexity was not our focus, we kept the game as simple as possible in this regards, disclaiming usual methods of making games challenging: time limits, an increasing number of tasks and goals to achieve... The principle task always remains the same (under varying conditions): the player has to steer both objects (e.g. two balls) into their target areas by tilting the MacBook. The task is fulfilled once both balls are in their target areas at the same time. When the task is fulfilled the target areas change their position and the game continues.

There are different strategies to reach the goal. Besides carefully balancing both balls one obvious strategy is to let them roll more or less randomly in circles until they by chance are in their target areas at the same time. By decreasing the size of the target areas gradually (e.g. every time a task is fulfilled) the second strategy becomes less effective and attentive careful balancing is necessary. On the other hand letting the balls circulate at the track demands a movement pattern which is interesting as well. Periodically tilting the MacBook in a rotatory manner and thus sustaining the velocity of the ball is a steering movement that clearly differs from a balancing task. It is here important to mention that objects react in different ways to inclinations due to different friction parameters whereby, with some skill, it is possible to let only one ball circulate and keep the other one fixed (more or less) at the same place. In contrast to balancing the ball, letting it rotate is a dynamical, periodical process. The player may therefore rely more on the dynamical information that he perceives aurally than on what can be seen at the screen. In this respect the "resonance-like" behaviour of the balls movement at periodic circular acceleration plays a decisive role. Of course the player can visually perceive in which way the ball reacts to a rotary tilting movement, but such kind of continuous, dynamical information is one strength of the auditive perceptual channel (compare e.g. [9]).

We finally thought about a way to force the player to perform balancing and "rotation control" simultaneously and designed the following scenario as part of the game: the size of the target area of one of two moving objects (on parallel circular tracks) is determined by the velocity of the second object. While the position of the target is constant its size increases with the velocity of the object moving on the parallel track. Under these conditions, in order to fulfill the task the ball that affects its neighbour's target size *must* move because a velocity of zero corresponds to a target size of zero. Letting said object rotate is now not the only way to reach the task but clearly (depending of course on the exact parameters of the coupling) the most obvious and the player is more or less forced to apply the following strategy:

- The faster the circulating ball moves the bigger becomes the target area of the other ball.
- If the ball with the varying target size reached it's goal the player has to keep it there.
- He would keep it there just for a moment, because the circulating ball frequently passes through it's goal.

In this scenario a rotary inclination of the device is a typical gesture the player performs to play the game — a circumstance that certainly supports our current intention (and work) of porting the application to a smaller, truly handheld device.

Altogether we have so far realized a number of instruments that may be combined: a set of sound models, two visually different objects, friction parameters to make them also "feel" different and targets that can vary in position and size (also velocity-dependent target-size). Of course the point is not to arbitrarily combine those building blocks which are all related to each other. A higher setting of the friction parameter will rather suit a sliding sound and the cylinder visualisation, just as the ball visualisation naturally fits the rolling sound. Under the aspect of creating a preferably convincing game one would try to set a combination that is as realistic as possible whereas for psychoacoustical tests it is also interesting to chose unintuitive (or even contradicting) combinations.

3 Technical Realisation / Example

We designed the game as a standalone application that should run on all laptops of the MacBook/MacBookPro series. The native application programming interface (API) for dealing with sound in Apple's Mac OS X operating system is Core Audio [3]. The audio processing of our program is implemented as a callback architecture with Core Audio[2], which means that Core Audio instructs our program to render audio output and write it into a buffer that can be referenced by an assigned memory address. It is important

---

[2] Here we use Core Audio's "Audio Unit Framework", a plug-in architecture that can be used for effects and virtual instruments.

to understand that we are not telling Core Audio to play back precasted audio samples, instead of that we are continuously computing frames of audio output ourselves and write them into an audio buffer.

Graphical rendering is done with OpenGL, a cross-platform API for computer graphics. For scheduling of in- and graphical output cross-platform free library "GLUT" is used. GLUT also supports callback driven event processing, which means that we also have a callback function for video processing (besides the function for audio).

3.1 Scheduling

When the application is started it registeres itself as a client for audio (Core Audio) and video processing (GLUT) and detects the motion sensor. All physicial simulation is computed with audio rate.

Regarding the physics there are basically two tasks to perform: calculate the movement of the object along the track and synthesise the sound. The rolling sound model is predicated on physical considerations of the interaction between ball and surface. Besides the macroscopic movement of the ball that can be seen there is a microscopic behaviour that can only be heard, both must be computed. One approach (actually the most common, in combination with sample-based audio) would be to use a physics engine for the calculation of the ball's macroscopic movement with a defined temporal resolution and synthesise audio feedback on the basis of the result. In this case the quality of the feedback would strongly depend on the update rate of the used physics engine. The opposite approach would be to unify macroscopic and microscopic movement in one model to obtain both as the output of the same algorithm, which would make movement and sound much more coherent but on the other hand is also much more difficult to achieve. In our simulation we chose a method which is a compromise of these two solutions. We make use of two separate models but both are always calculated in parallel within the same cycle of the audio processing, which means that we update the position of the ball with audio rate and instantaneously calculate the next sample frame of audio feedback in dependence of the position and velocity parallel to the track.

Structure and scheduling are as follows: all physical objects and algorithms of the current active simulation are contained in a data structure that we call "scene". In fact we have a number of different scenes with different objects, sounds, etc., but only one scene is active at a time (the others are stored in memory, ready to be activated). Only the currently active scene is updated, which means that the physical behaviour of the objects and the sound output is computed (with audio rate). These updates are performed within the audio callback function. The video callback function induces the rendering of the scene by accessing relevant data computed in the audio loop.

The buffer size of our audio callback routine is fixed at 1024 samples, which means that the audio callback function is called with about $43 Hz$ (at an audio rate of $44100 Hz$). Each time the audio callback function is executed the accelerometer data is read *once*, *1024* updates of the scene are performed and the buffer is filled with samples. We have no direct control about the exact moment when the audio callback function is called, we only know that it happens in average 43 times per second. By relying on the callback cycle of the audio driver a certain jitter is unavoidable which we however accept since an average update rate of $43Hz$ is quite high for accelerometer data used for balancing control.

As already mentioned the video rendering is also executed by means of a callback function, for which we chose a fixed rate of $50Hz$. Since values relevant for graphical output (object position...) are computed in the audio callback cycle the same remarks apply here as for updates of accelerometer data.

3.2 Sound Feedback

In our game sound feedback is not just a fancy supplement but one of our central interests. We seek to explore the effects of the sound on the player. Does it enhance his performance? Does it affect his attitude, his perception of the game? Recent experiments with the *Ballancer* interface [8] showed that the perception of velocity plays a key role in this context, it can improve a player's performance in a control task [9].

As already mentioned our game is based on steering objects on plains affected by gravity. Hence the scenario of a rolling ball is the starting point of our considerations about *sound models* for the game, about their qualities and their implementation. The "rolling sound model" considers the profile and the vibration of the surface as well as the vertical interaction of the surface and the rolling ball. It is our informal supposition that the detailed nature of this vertical interaction plays a strong role in the auditory perception of rolling: it appears that an important factor for sounds to be classified by human listeners as "rolling" is the mixture of short periods of microscopic bouncing and longer periods of contact between the rolling object and the surface. The "sliding sound model" only considers the vibration of the surface and the sweeping of the object across the surface profile. In accordance with our informal supposition on the characteristics of contact sound the audible result resembles the sound of a sliding object.

In our scenario two objects are interacting with each other: the ball (resp. the cylinder) and the surface. There are different approaches to describe the behaviour of vibrating and interacting objects; we chose the modal approach (compare [7]) for a model of the inner resonance behaviour of the vibrating surface. We did that on the assumption that the surface's vibration plays the dominant part in the sound that we hear when an object is rolling or sliding across it while the vibration of the solid rolling object itself (e.g. a marble) can hardly be heard directly. The modal description of the surface (in the following "modal object") is the source where our sound output comes from: the velocity of the surface is written into the audio buffer.

If an impulsive force is applied to the modal object is starts vibrating and becomes silent again after some time due to inner friction. As the ball is bouncing it applies a force to the surface. A rolling ball or a sliding object is continuously impacting the surface because natural surfaces are not perfectly smooth. There are different approaches to model that contact. Regarding the profile as a (force) input signal to the modal object is one way, discussed in the next subsection (sliding sound). A more accurate approach is to include physical and geometrical considerations: the rolling sound model.

3.2.1 Sliding Sound

This sound model is based upon assumptions about the impact force that the sliding object applies to the surface: the impact force has a noise-like character and is bandlimited in the frequency domain. In other words an incompressible object perfectly strobes the surface's noise-like outline. Vertical interaction between object and surface is not considered, as well as the geometry of the object (e.g. the ball's shape).

Hence we utilise the profile of the surface as (force) input signal to the modal object (in a way similar to some previously used algorithms, compare e.g. [10]). We use bandpass filtered white noise as signal. The centre frequency of the filter is proportional to the (horizontal) velocity of the ball, the bandwidth is kept constant; the sound model can be regarded as a modal resonator.

3.2.2   Rolling Sound

The rolling sound model affords a high degree of realism, it is the natural acoustic complement of what can be seen on the screen when the balls are rolling at their tracks. The rolling sound model embraces the vibrations of the surface, the surface's profile and the vertical contact interaction of the vibrating surface and the ball. Details of the algorithm are described in dedicated articles [6] [7], in the following we try to give an idea of the main points.

The basic principle of the rolling sound has already been introduced in [6] and implemented as a patch for Pure Data[3]. Recently an improved contact model has been presented [7]. In the present work we implement the improved algorithm of [6] based upon the contact model of [7] implemented in C++. In the following we give a short summary about the ideas and methods.

Despite some interesting psychoacoustic studies (e.g. [2]) the question of how to exactly describe the acoustic features responsible for a sound event to be perceived as "rolling" is still not perfectly answered. This observation supports a physics-based approach in the synthesis of rolling sounds. We look at "rolling" as a process of sustained bouncing, eventually microscopic bouncing. The second aspect is that we allow the ball to penetrate the surface, thus a contact is not a change of state at a discrete moment in time. Contacts are continuing processes that have to be modelled. To sum up, the interaction can be both continuous and sporadic: there are moments of contact, in particular continuous contact, and moments when surface and ball are not in contact.

The modelling of the contact is one key building block of the rolling sound model. [7] introduces a method of energy-stable modelling of continuous or repeated contact in dynamical situations, a method that is also applied here. The basic idea is to use a modal description for the period of contact too. If there is no contact the state of the surface is described by a modal object, the ball's state is stored separately. In case of contact surface and ball are regarded as one object, which is described by means of modal parameters as well. The model of rolling includes frequent "switching" between both configurations.

Besides physical considerations (improving the contact model) there are also geometric aspects to consider. Although we constrain that the contact only takes place at one contact point the ball of course has a geometric attribute that should not be disregarded: its radius. We simplify the profile of the surface to a white noise distribution of profile points (with constant horizontal distance, but scattered vertical position). One

---

[3] A graphical programming language for the creation of interactive computer music and multimedia works.

can imagine that the ball cannot touch every point since it is too big to fill each gap. Therefore the profile needs to be filtered. Instead of bandpass filtering (as in the next section) we have to filter the "profile signal" based on geometrical considerations. The exact procedure is explained in [6] and for now called "rolling filter". To sum up, the result is a profile signal consisting of arcs of circles which is the path that a ball would follow if it was perfectly succeeding the surface's profile.

In the sliding sound model the profile signal is used directly as input force for the modal object. Since the rolling sound model contains the modelling of contacts (and resulting forces) based on physical considerations the profile is regarded as a position-dependent (and thus timevariant) input parameter to the contact model. The contact model described above involves a continuous computation of the distance between ball and surface. At this point the profile of the surface is incorporated as a vertical distance offset. Besides several "white noise variants" as profiles for surfaces the rolling filter technique also enables more complex profiles like saw tooth surfaces[4] that enable interesting possibilities to imitate natural surfaces.

4 Summary and Future Work

We have described a recently implemented game on the MacBook platform intended for use as easily distributable tool for collection of psychophysical data. The software uses the MacBook's built-in accelerometer for gestural control of a balancing scenario. It offers precise synchronous visual and auditory feedback based on models of different degrees of complexity of physical contact. A first demo-version of the game platform — fully functional in its main components — is ready for free download and installation on recent generation MacBooks.

Of course this contribution is to be understood as a report on the development and potential of the presented software as basis for future online data acquisition. Concrete experiments at the platform are currently being planned. To this end, mechanisms of automated online data collection are currently being incorporated into a next version of the software and concrete experimental game concepts are being developed. Also, the plan of porting the environment to the iPhone, a platform that might further enhance possibilities of distribution and user interaction is being explored more closely.

References

1. Carello, C., Anderson, K.L., Kunkler-Peck, A.J.: Perception of object length by sound. Psychological Science 9(3), 211–214 (1998)
2. Houben, M., Kohlrausch, A., Hermes, D.: Auditory cues determining the perception of size and speed of rolling balls. In: ICAD 2001, Espoo, Finland, pp. 105–110 (2001)
3. Apple Inc. (2009),
   http://developer.apple.com/documentation/musicaudio
4. Klatzky, R.L., Pai, D.K., Krotkov, E.P.: Perception of material from contact sounds. Presence: Teleoperators and Virtual Environment 9(4), 399–410 (2000)
5. Lederman, S.J.: Auditory texture perception. Perception 8, 93–103 (1979)

---

[4] Combined with above-mentioned white noise "micro structure"...

6. Rath, M.: An expressive real-time sound model of rolling. In: Proceedings of the 6th International Conference on Digital Audio Effects(DAFx 2003), London, United Kingdom (September 2003)
7. Rath, M.: Energy-stable modelling of contacting modal objects with piece-wise linear interaction force. In: Proceedings of the 11th International Conference on Digital Audio Effects (DAFx 2008), Espoo, Finland (September 2008)
8. Rath, M., Rocchesso, D.: Informative sonic feedback for continuous human–machine interaction — controlling a sound model of a rolling ball. IEEE Multimedia Special on Interactive Sonification 12(2), 60–69 (2005)
9. Rath, M., Schleicher, R.: On the relevance of auditory feedback for quality of control in a balancing task. Acta Acustica United With Acustica 94(1), 12–20 (2008)
10. van den Doel, K., Kry, P.G., Pai, D.K.: Foleyautomatic: Physically-based sound effects for interactive simulation and animation. In: Proc. ACM Siggraph 2001, Los Angeles (August 2001)
11. Wildes, R.P., Richards, W.A.: Recovering material properties from sound. Natural Computation, 356–363 (1988)

Author Index